SOME ENGLISH GARDENS

Gertrude Jekyll

Table of Contents

PREFACE

THE publication of this collection of reproductions of water-colour drawings would have been impossible without the willing co-operation of the owners of the originals. Special acknowledgment is therefore due to them for their kindness and courtesy, both in consenting to such reproduction and in sparing the pictures from their walls. On pages xi. and xii. is given a full list of the pictures, together with the names of the owners to whom we are so greatly indebted.

We have also had the valuable assistance of Mr. Marcus B. Huish, of The Fine Art Society, who has taken the greatest interest in the work from its inception.

G. S. E.

G. J.

BROCKENHURST

THE English gardens in which Mr. Elgood delights to paint are for the most part those that have come to us through the influence of the Italian Renaissance; those that in common speech we call gardens of formal design. The remote forefathers of these gardens of Italy, now so well known to travellers, were the old pleasure-grounds of Rome and the neighbouring districts, built and planted some sixteen hundred years ago.

Though many relics of domestic architecture remain to remind us that Britain was once a Roman colony, and though it is reasonable to suppose that the conquerors brought their ways of gardening with them as well as their ways of building, yet nothing remains in England of any Roman gardening of any importance, and we may well conclude that our gardens of formal design came to us from Italy, inspired by those of the Renaissance, though often modified by French influence.

Very little gardening, such as we now know it, was done in England earlier than the sixteenth century. Before that, the houses of the better class were places of defence; castles, closely encompassed with wall or moat; the little cultivation within their narrow bounds being only for food—none for the pleasure of garden beauty.

But when the country settled down into a peaceful state, and men could dwell in safety, the great houses that arose were no longer fortresses, but beautiful homes both within and without, inclosing large garden spaces, walled with brick or stone only for defence from wild animals, and divided or encompassed with living hedges of yew or holly or hornbeam, to break

wild winds and to gather on their sunny sides the life-giving rays that flowers love.

So grew into life and shape some of the great gardens that still remain; in the best of them, the old Italian traditions modified by gradual and insensible evolution into what has become an English style. For it is significant to observe that in some cases, where a classical model has been too rigidly followed, or its principles too closely adhered to, that the result is a thing that remains exotic—that will not assimilate with the natural conditions of our climate and landscape. What is right and fitting in Italy is not necessarily right in England. The general principles may be imported, and may grow into something absolutely right, but they cannot be compelled or coerced into fitness, any more than we can take the myrtles and lentisks of the Mediterranean region and expect them to grow on our middle-England hill-sides. This is so much the case, with what one may call the temperament of a region and climate, that even within the small geographical area of our islands, the comparative suitability of the more distinctly Italian style may be clearly perceived, for on our southern coasts it is much more possible than in the much colder and bleaker midlands.

Thus we find that one of the best of the rather nearly Italian gardens is at Brockenhurst in the New Forest, not far from the warm waters of the Solent. The garden, in its present state, was laid out by the late Mr. John Morant, one of a long line of the same name owning this forest property. He had absorbed the spirit of the pure Italian gardens, and his fine taste knew how to bring it forth again, and place it with a sure hand on English soil.

It is none the less beautiful because it is a garden almost without flowers, so important and satisfying are its permanent forms of living green walls, with their own proper enrichment of ball and spire, bracket and buttress, and so fine is the design of the actual masonry and sculpture.

The large rectangular pool, known as the Canal, bordered with a bold kerb, has at its upper end a double stair-way; the retaining wall at the head of the basin is cunningly wrought into buttress and niche. Every niche has its appropriate sculpture and each buttress-pier its urn-like finial. On the upper level is a circular fountain bordered by the same kerb in lesser proportion, with stone vases on its circumference. The broad walk on both levels is bounded by close walls of living

THE TERRACE, BROCKENHURST
FROM THE PICTURE IN THE POSSESSION OF Mr. G. N. Stevens

greenery; on the upper level swinging round in a half circle, in which are cut arched niches. In each leafy niche is a bust of a Cæsar in marble on a tall term-shaped pedestal. Orange trees in tubs stand by the sides of the Canal. This is the most ornate portion of the garden, but its whole extent is designed with equal care. There is a wide bowling-green for quiet play; turf walks within walls of living green; everywhere that feeling of repose and ease of mind and satisfaction that comes of good balance and proportion. It shows the classical sentiment thoroughly assimilated, and a judicious interpretation of it brought forth in a form not only possible but eminently successful, as a garden of Italy translated into the soil of one of our Southern Counties.

Whether or not it is in itself the kind of gardening best suited for England may be open to doubt, but at least it is the work of a man who knew what he wanted and did it as well as it could possibly be done. Throughout it bears evidence of the work of a master. There is no doubt, no ambiguity as to what is intended. The strong will orders, the docile stone and vegetation obey. It is full-dress gardening, stately, princely, full of dignity; gardening that has the courtly sentiment. It seems to demand that the actual working of it should be kept out of sight. Whereas in a homely garden it is pleasant to see people at work, and their tools and implements ready to their hands, here there must be no visible intrusion of wheelbarrow or shirt-sleeved labour.

Possibly the sentiment of a garden for state alone was the more gratifying to its owner because of the near neighbourhood of miles upon miles of wild, free forest; land of the same character being inclosed within the

property; the tall trees showing above the outer hedges and playing to the lightest airs of wind in an almost strange contrast to the inflexible green boundaries of the ordered garden.

The danger that awaits such a garden, now just coming to its early prime, is that the careful hand should be relaxed. It is an heritage that carries with it much responsibility; moreover, it would be ruined by the addition of any commonplace gardening. Winter and summer it is nearly complete in itself; only in summer flowers show as brilliant jewels in its marble vases and in its one restricted parterre of box-edged beds.

It is a place whose design must always dominate the personal wishes, should they desire other expression, of the succeeding owners. The borders of hardy and half-hardy plants, that in nine gardens out of ten present the most obvious ways of enjoying the beauty of flowers, are here out of place. In some rare cases it might not be impossible to introduce some beautiful climbing plant or plant of other habit, that would be in right harmony with the design, but it should only be attempted by an artist who has such knowledge of, and sympathy with, refined architecture as will be sure to guide him aright, and such a consummate knowledge of plants as will at once present to his mind the identity of the only possible plants that could so be used. Any mistaken choice or introduction of unsuitable plants would grievously mar the design and would introduce an element of jarring incongruity such as might easily be debased into vulgarity.

There is no reason why such other gardening may not be rightly done even at Brockenhurst, but it should not encroach upon or be mixed up with an Italian design. Its place would be in quite another portion of the grounds.

BROCKENHURST: THE GARDEN GATE
FROM THE PICTURE IN THE POSSESSION OF MISS RADCLIFFE

HOLLYHOCKS AT BLYBOROUGH

THE climate of North Lincolnshire is by no means one of the most favourable of our islands, but the good gardener accepts the conditions of the place, faces the obstacles, fights the difficulties, and conquers.

Here is a large walled garden, originally all kitchen garden; the length equal to twice the breadth, divided in the middle to form two squares. It is further subdivided in the usual manner with walks parallel to the walls, some ten feet away from them, and other walks across and across each square. The paths are box-edged and bordered on each side with fine groups of hardy flowers, such as the Hollyhocks and other flowers in the picture.

The time is August, and these grand flowers are at their fullest bloom. They are the best type of Hollyhock too, with the wide outer petal, and the middle of the flower not too tightly packed.

Hollyhocks have so long been favourite flowers—and, indeed, what would our late summer and autumn gardens be without them?—that they are among those that have received the special attention of raisers, and have become what are known as florists' flowers. But the florists' notions do not always make for the highest kind of beauty. They are apt to favour forms that one cannot but think have for their aim, in many cases, an ideal that is a false and unworthy one. In the case of the Hollyhock, according to the florist's standard of beauty and correct form, the wide outer petal is not to be allowed; the flower must be very tight and very round. Happily we need not all be florists of this narrow school, and we are at liberty to

try for the very highest and truest beauty in our flowers, rather than for set rules and arbitrary points of such extremely doubtful value.

The loosely-folded inner petals of the loveliest Hollyhocks invite a wonderful play and brilliancy of colour. Some of the colour is transmitted through the half-transparency of the petal's structure, some is reflected from the neighbouring folds; the light striking back and forth with infinitely beautiful trick and playful variation, so that some inner regions of the heart of a rosy flower, obeying the mysterious agencies of sunlight, texture and local colour, may tell upon the eye as pure scarlet; while the wide outer petal, in itself generally rather lighter in colour, with its slightly waved surface and gently frilled edge, plays the game of give and take with light and tint in quite other, but always delightful, ways.

Then see how well the groups have been placed; the rosy group leading to the fuller red, with a distant sulphur-coloured gathering at the far end; its tall spires of bloom shooting up and telling well against the distant tree masses above the wall. And how pleasantly the colour of the rosy group is repeated in the Phlox in the opposite border. And what a capital group that is, near the Hollyhocks of that fine summer flower, the double Crown Daisy (*Chrysanthemum coronarium*), with the bright glimpses of some more of it beyond. Then the Pansies and Erigerons give a mellowing of grey-lilac that helps the brighter colours, and is not overdone.

The large fruit-tree has too spreading a shade to allow of much actual bloom immediately beneath it, so that here is a patch of Butcher's Broom, a shade-loving plant. Beyond, out in the sunlight again, is the fine

herbaceous Clematis (*C. recta*), whose excellent qualities entitle it to a much more frequent use in gardens.

The flower-borders are so full and luxuriant that they completely hide the vegetable quarters within, for the garden is still a kitchen garden as to its main inner spaces. These masses of good flowers are the work of the Misses Freeling; they are ardent gardeners, sparing themselves no labour or trouble; to their care and fine perception of the best use of flowers the beauty and interest of these fine borders are entirely due. Indeed, this garden is a striking instance

BLYBOROUGH: HOLLYHOCKS
FROM THE PICTURE IN THE POSSESSION OF Mr. C. E.
Freeling

of the extreme value of personal effort combined with knowledge and good taste.

These qualities may operate in different gardens in a hundred varying ways, but where they exist there will be, in some form or other, a delightful garden. Endless are the possibilities of beautiful combinations of flowers; just as endless is their power of giving happiness and the very purest of human delight. So also the special interest of different gardens that are personally directed by owners of knowledge and fine taste would seem to be endless too, for each will impress upon it some visible issue of his own perception or discernment of beauty.

About the house and lawns are other beds and borders of herbaceous flowers of good grouping and fine growth; conspicuous among them is that excellent flower *Campanula pyramidalis*, splendidly grown.

Though Blyborough is in a cold district, it has the advantage of lying well sheltered below a sharply-rising ridge of higher land.

GREAT TANGLEY MANOR

FORTY years ago, lying lost up a narrow lane that joined a track across a wide green common, this ancient timber-built manor-house could scarcely have been found but by some one who knew the country and its by-ways well. Even when quite near, it had to be searched for, so much was it hidden away behind ricks and farm-buildings; with the closer overgrowth of old fruit trees, wild thorns and elders, and the tangled wastes of vegetation that had invaded the outskirts of the neglected, or at any rate very roughly-kept, garden of the farm-house, which purpose it then served.

What had been the moat could hardly be traced as a continuous water-course; the banks were broken down and over-grown, water stood in pools here and there; tall grass, tussocks of sedge and the rank weeds that thrive in marshy places had it all to themselves.

But the place was beautiful, for all the neglect and disorder, and to the mind of a young girl that already harboured some appreciative perception of the value of the fine old country buildings, and whose home lay in a valley only three miles away, Tangley was one of the places within an easy ride that could best minister to that vague unreasoning delight, so gladly absorbed and so keenly enjoyed by an eager and still almost childish imagination. For the mysteries of romantic legend and old tale still clung about the place—stories of an even more ancient dwelling than this one of the sixteenth century.

There was always a ready welcome from the kindly farmer's wife, and complete freedom to roam about; the pony was accommodated in a cowstall, and many happy summer hours were spent in the delightful

THE PERGOLA, GREAT TANGLEY
FROM THE PICTURE IN THE POSSESSION OF MR. WICKHAM
FLOWER

wilderness, with its jewel of a beautifully-wrought timbered dwelling that had already stood for three hundred years.

In later days, when the whole of the Grantley property in the district was sold, Great Tangley came into the market. Happily, it fell into the best of hands, those of Mr. and Mrs. Wickham Flower, and could not have been better dealt with in the way of necessary restoration and judicious addition.

The moat is now a clear moat again; and good modern gardening, that joins hands so happily with such a beautiful old building, surrounds it on all sides. There was no flower garden when the old place was taken in hand; the only things worth preserving being some of the old orchard trees within the moat to the west. A space in front of the house, on its southward face, inclosed by loop-holed walls of considerable thickness, was probably the ancient garden, and has now returned to its former use. The modern garden extends over several acres to the east and south beyond the moat. The moat is fed by a long-shaped pond near its south-eastern angle. The water margin is now a paradise for flower-lovers, with its masses of water Irises and many other beautiful aquatic and sub-aquatic plants; while Water-Lilies, and, surprising to many, great groups rising strongly from the water of the white Calla, commonly called Arum Lily, give the pond a quite unusual interest. To the left is an admirable bog-garden with many a good damp-loving plant, and, best of all in their flowering time, some glorious clumps of the Moccasin Flower (*Cypripedium spectabile*), largest, brightest, and most beautiful of hardy orchids.

Those who have had the luck to see this grand plant at Tangley, two feet high and a mass of bloom, can understand the admiration of others who have met with it in its North American home, and their description of how surprisingly beautiful it is when seen rising, with its large rose and white flowers, and fresh green pleated leaves, from the pools of black peaty mud of the forest openings. But it seems scarcely possible that it can be finer in its own home than it is in this good garden.

19

Beyond the bog-garden, on drier ground, is a garden of heaths, and, returning by the pathway on the other side of the pond, is the kitchen garden, a strip of pleasure-ground being reserved between it and the pond. Here is the subject of the picture. The pergola runs parallel with the pond, which, with the house and inclosed garden, are to the spectator's right. To the left, before the vegetable quarters begin, is a capital rock-garden of the best and simplest form—just one long dell, whose sides are set with rocks of the local Bargate stone and large sheets of creeping and rock-loving plants. Taller green growths of shrubby character shut it off from other portions of the grounds.

The picture speaks for itself. It tells of the right appreciation of the use of the good autumn flowers, in masses large enough to show what the flowers will do for us at their best, but not so large as to become wearisome or monotonous. Roses, Vines and Ivies cover the pergola, making a grateful shade in summer. Each open space to the right gives a picture of water and water-plants with garden ground beyond, and, looking a little forward, the picture is varied by the background of roof-mass with a glimpse of the timbered gables of the old house.

The new garden is growing mature. The Yews that stand like gate-towers flanking the entrance of the green covered way, have grown to their allotted height, doing their duty also as quiet background to the autumnal flower-masses. In the border to the left are Michaelmas Daisies, French Marigolds, and a lower growth of Stocks; to the right is a dominating mass of the great white Pyrethrum, grouped with pink Japan Anemone, Veronicas and yellow Snapdragon. Japan Anemones, both pink and white,

are things of uncertain growth in many gardens of drier soil, but here, in the rich alluvial loam of a valley level, they attain their fullest growth and beauty.

BULWICK: AUTUMN

FROM THE PICTURE IN THE POSSESSION OF Lord Henry

Grosvenor

BULWICK HALL

BULWICK HALL, in Northamptonshire, the home of the Tryon family, but, when the pictures were painted, in the occupation of Lord and Lady Henry Grosvenor, is a roomy, comfortable stone building of the seventeenth century. The long, low, rather plain-looking house of two stories only, is entered in an original manner by a doorway in the middle of a stone passage, at right angles to the building, and connecting it with a garden house. The careful classical design and balustraded parapet of the outer wall of this entrance, and the repetition of the same, only with arched openings, to the garden side, scarcely prepare one for the unadorned house-front; but the whole is full of a quiet, simple dignity that is extremely restful and pleasing. Other surprises of the same character await one in further portions of the garden.

Passing straight through the entrance gate there is a quiet space of grass; a level court with flagged paths, bounded on the north by the house and on the east and west by the arcade and the wall of the kitchen garden. The ground falls slightly southward, and the fourth side leads down to the next level by grass slopes and a flight of curved steps widening below. Trees and shrubs are against the continuing walls to right and left, and beds and herbaceous borders are upon the grassy space. The wide green walk, between long borders of hardy plants, leading forward from the foot of the steps, reaches a flower-bordered terrace wall, and passes through it by a stone landing to steps to right and left on its further side. A few steps descend in twin flights to other landings, from which a fresh flight on each side reaches the lowest garden level, some nine feet below the last. The

whole of this progression, with its pleasant variety of surface treatment and means of descent, is in one direct line from a garden door in the middle of the house front.

The lowest flight of steps, the subject of the first picture, has a simple but excellent wrought-iron railing, of that refined character common to the time of its making. It was draped, perhaps rather over-draped when the picture was painted, with a glory of Virginia Creeper in fullest gorgeousness of autumn colouring. This question of the degree to which it is desirable to allow climbing plants to cover architectural forms, is one that should be always carefully considered. Bad architecture abounds throughout the country, and free-growing plants often play an entirely beneficent part in concealing its mean or vulgar or otherwise unsightly character. But where architectural design is good and pure, as it is at Bulwick, care should be taken in order to prevent its being unduly covered. Old brick chimney-stacks of great beauty are often smothered with Ivy, and the same insidious native has obliterated many a beautiful gate-pier and panelled wall. But the worst offender in modern days has been the far-spreading Ampelopsis Veitchii, useful for the covering of mean or featureless buildings, but grievously and mischievously out of place when, for instance, ramping unchecked over the old brickwork of Wolsey's Palace at Hampton Court. Some may say that it is easily pulled off; but this is not so, for it leaves behind, tightly clinging to the old brick surface, the dried-up sucker and its tentacle, desiccated to a consistency like iron wire. These are impossible to detach without abrasion of surface,

24

while, if left, they show upon the brick as a scurfy eruption, as disfiguring to the wall-face as are the scars of smallpox on a human countenance.

The iron-railed steps in the picture come down upon a grassy space rather near its end. Behind the spectator it stretches away for quite four times the length seen in the picture. It is bounded on the side opposite the steps by a long rectangular fish-pond. The whole length of this is not seen, for the grass walk narrows and passes between old yew hedges, one on the side of the pond, the other backed by some other trees against the kitchen garden wall, which is a prolongation of the terrace wall in the picture.

The garden is still beautifully kept, but owes much of its wealth of

BULWICK: THE GATEWAY
FROM THE PICTURE IN THE POSSESSION OF LORD HENRY
GROSVENOR

hardy flowers to the planting of Lady Henry Grosvenor, whose fine taste and great love of flowers made it in her day one of the best gardens of hardy plants, and whose untimely death, in the very prime of life, was almost as much deplored by the best of the horticultural amateurs who only knew her by reputation, but were aware of her good work in gardening, as by her wide circle of personal friends.

She had a special love for the flag-leaved Irises, and used them with very fine effect. The borders that show to right and left of the steps had them in large groups, and were masses of bloom in June; other plants, placed behind and between, succeeding them later. Lady Henry was one of the first amateurs to perceive the value of planting in this large way, and, as she had ample spaces to deal with, the effects she produced were very fine, and must have been helpful in influencing horticultural taste in a right direction.

Another important portion of the garden at Bulwick is a long double flower-border backed by holly hedges, that runs through the whole middle length of the kitchen garden. It is in a straight line with the flagged walk that passes westward across the green court next to the house, and parallel with its garden front. The flagged path stops at the gate-piers in the second picture, a grass path following upon the same line and passing just behind the shaded seat.

The holly hedges that back the borders are old and solid. Their top line, shaped like a flat-pitched roof, is ornamented at intervals with mushroom-shaped finials, each upon its stalk of holly stem. The grass walk and double border pass right across the kitchen garden in the line of its longest axis.

At the furthest end there is another pair of the same handsome gate-piers with a beautiful wrought-iron gate, leading into the park. The park is handsomely timbered, and in early summer is especially delightful from the great number of fine old hawthorns.

In Lady Henry's time several borders in the kitchen garden were made bright with annuals and other flowers. Such borders are very commonly used for reserve purposes, such as the provision of flowers for cutting, with one main double border for ornament alone. But where gardens are being laid out from the beginning, such a plan as this at Bulwick, of a grass path with flower borders and a screening hedge at the back, passing through a kitchen garden, is an excellent one, greatly enlarging the length of view of the pleasure garden, while occupying only a relatively small area. It is also well in planning a garden to provide a reserve space for cutting alone, of beds four feet, and paths two feet wide, and of any length suitable for the supply required. This has the advantage of leaving the kitchen garden unencumbered with any flower-gardening, and therefore more easy to work.

Such a long-shaped garden is also capable of various ways of treatment as to its edge, which need not necessarily be an unbroken line. The length of the border in question is perhaps a little too great. It might be better, while keeping the effect of a quiet line, looking from end to end, to have swung the edge of the border back in a segment of a circle to a little more than half its depth, every few yards, in such a proportion as a plan to scale would show to be right; or to have treated it in some one of the many possible ways of accentuation where the cross paths occur that divide it

into three lengths. The thinking out of these details according to the conditions of the site, the combining of them into designs that shall add to its beauty, and the actual working of them, the mind meanwhile picturing the effect in advance—these are some of the most interesting and enlivening of the many kinds of happiness that a garden gives.

Be it large or small there is always scope for inventive ability; either for the bettering of something or for the casting of some detail into a more desirable form. Every year brings some new need; in supplying it fresh experience is gained, and with this an increasing power of adapting simple means to such ends as may be easily devised to the advancement of the garden's beauty.

BRAMHAM

THE gardens at Bramham in Yorkshire, laid out and built near the end of the seventeenth century, are probably the best preserved in England or the grounds that were designed at that time under French influence. Wrest in Bedfordshire, and Melbourne in Derbyshire of which some pictures will follow, are also gardens of purely French character.

It is extremely interesting to compare these gardens with those of a more distinctly Italian feeling. Many features they have in common; architectural structure and ornament, close-clipped evergreen hedges inclosing groves of free-growing trees; parterres, pools and fountains. Yet the treatment was distinctly different, and, though not easy to define in words, is at once recognised by the eye.

For one thing the French school, shown in its extremest form by the gardens of Versailles, dealt with much larger and more level spaces. The gardens of Italian villas, whether of the Roman Empire or of the Renaissance, were for the most part in hilly places; pleasant for summer coolness. This naturally led to much building of balustraded terraces and flights of steps, and of parterres whose width was limited to that of the level that could conveniently be obtained. Whereas in France, and in England especially, where the country house is the home for all the year, the greater number of large places have land about them that is more or less level and that can be taken in to any extent.

At Bramham the changes of level are not considerable, but enough to furnish the designer with motives for the details of his plan. The house, of about the same date as the garden, was internally destroyed by fire in the

last century. The well-built stone walls still stand, but the building has never been restored. The stables and kennels are still in use, but the owner, Captain Lane-Fox, lives in another house on the outskirts of the park. The design of the gardens has often been attributed to Le Nôtre, and is undoubtedly the work of his school, but there is nothing to prove that the great French master was ever in England.

The way to the house is through a large, well-timbered park. Handsome gate-piers with stone-wrought armorial ornament lead into a forecourt stretching wide to right and left. A double curved stairway ascends to the main door. To the left of the house is an entrance to the garden through a colonnade. Next to the garden front of the house, which faces south-west, is a broad gravelled terrace. The ground rises away from the house by a gently sloping lawn, but in the midmost space is a feature that is frequent in the French gardening of the time, though unusual in England: a long theatre-shaped extent of grass. There is a stone sundial standing on two wide steps near the house, and a gradually heightened retaining wall following the rise of the ground. Not more than two feet high where it begins below, and there accentuated on either side by a noble stone plinth and massive urn, the retaining wall, itself a handsome object of bold masonry, follows a straight line for some distance, and then swings round in a segmental curve to meet the equal wall on the further side; thus inclosing a space of level sward. Midway in the curve, where the wall is some twelve feet high, there appear to have been niches in the masonry, possibly for fountains.

The wide gravel walk next the house-front falls a little as it passes to the left, divides in two and continues by an upward slope on either side of a wall-fountain in a small inclosure formed by the retaining walls of the rising paths. The path then passes all round the large rectangular pool, one end of which forms the subject of the picture. This shows well the graceful ease and, one may say, the courteous suavity, that is the foremost character of this beautiful kind of French designing. The high level of the water in the pool, so necessary for good effect, is a detail that is often overlooked in English gardens. Nothing looks worse than a height of bare wall in a pool or fountain basin, and nothing is more commonly seen in our gardens. The low stone kerb bordering the pool is broken at intervals with only slightly rising pedestals for

THE POOL, BRAMHAM
FROM THE PICTURE IN THE POSSESSION OF
SIR JAMES WHITEHEAD, BART.

31

flower vases. Tubs of Agapanthus stand on the projections by the side of the piers that flank the small fountain basin, whose overflow falls into the pool.

All this portion of the garden has a background of yew hedges inclosing large trees. From this pool the ground rises to another; also of rectangular form, but with an arm to the right, in the line of the cross axis, forming a T-shape. Between the two, on a path always rising by occasional flights of steps, is a summer-house. The path swings round it in a circle. To right and left are flower-beds and roses; outside these, also on a curved line, are ranged a series of gracefully sculptured *amorini*, bearing aloft vases of flowers.

The path soon reaches the upper pool, again passing all round it. At the point furthest to the right, at the end of the projecting arm, and looking along the cross axis to where, beyond the pool, the ground again rises, is a handsome wall fountain, with steps to right and left, inclosed by panelled walls. All this garden of pool and fountain, easy way of step and gravel, and ornament of flower and sculpture, is bounded by the massive walls of yew, and all beyond is sheltering quietude of ancient trees. From several points around the highest pool, as well as from the rising lawns to right and left of the theatre, straight grass-edged paths, bordered by clipped hornbeam, lead through the heavily wooded ground. From distant points the main walks converge; and here, in a circular green-walled court, stands a tall pedestal bearing a handsome stone vase. The prospects down the alleys are variously ended; some by pillared temples set in green niches, some by the open park-landscape; some by further depths of woodland. It

is all easy and gracious, but full of dignity—courtly—palatial; bringing to mind the stately bearing and refined courtesy of manner of our ancestors of two centuries ago. It is good to know that some of these gardens and disciplined woodlands still exist in our own land and in France; these quiet *bosquets de verdure* of those far-away days. Though the scale on which they were planned is only suitable for the largest houses and for wealthy owners who can command lavish employment of labour, yet we cannot but admire the genius of those garden artists of France who designed so boldly and yet so gracefully, and who have left us such admirable records of their abounding ability.

MELBOURNE

THE gardens of Melbourne Hall in Derbyshire, the property of Earl Cowper, but occupied for the last five-and-twenty years by Mr. W. D. Fane, though perhaps less well preserved than those of Bramham, still show the design of Henry Wise in the early years of the eighteenth century. There had formerly been an older garden. Wise's plan shows how completely the French ideas had been adopted in England, for here again are the handsome pools and fountains, the garden thick-hedged with yew, and the *bosquet* with its straight paths, green-walled, leading to a large fountain-centred circle in the thickest of the grove.

The whole space occupied by the house and grounds is not of great extent; it is irregular and even awkward in shape, and has roads on two sides.

The treatment is extremely ingenious; indeed, it is doubtful whether any other plan that could have been devised would have made so much of the space or could have so cleverly concealed the limits.

The garden lies out forward of the house in a long parallelogram. Next to the house-front is the usual wide gravel terrace, from which paths, inclosing spaces of lawn, lead down to a lower level. The whole lawn, with its accompanying paths, slopes downward; where a steeper slope occurs above and below, the path becomes a flight of steps.

The lower level is intersected by paths. As they converge, they swing round the pedestal of the Flying Mercury that stands upon a circular grass-plot. The main path soon reaches the edge of the handsome pool known as the Great Water. It is four-sided, with a

MELBOURNE
FROM THE PICTURE IN THE POSSESSION OF
Mr. W. V. R. Fane

further semi-circular bay. A wide grass verge and turf slope form the edging. Broad walks pass all round, with pleasant views at various points into the cool and shaded woodland alleys. Near the further angles of the pool's green court, the great yew hedge, which bounds the whole garden, swings back into shallow segmental niches to take curved stone seats. Just beyond, on the return angle, the view from the path, here passing the right side of the pool, is ended by the lead figure of Perseus, of heroic size, also

standing in a niche cut in the yews. The companion statue of Andromeda occupies the corresponding niche on the other side.

After passing the Mercury, the view across the pool is met by a curious piece of wrought-iron work in the form of a high, dome-topped summer-house; a masterpiece of Jean Tijou. It is entered by steps, and leads, through the trees, to higher ground beyond.

Right and left of the middle and upper portions of the garden the great yew hedges are double; planted in parallel lines, with an open space between. Scotch Firs, now very old and towering high aloft, give great character to this part of the garden. In one place there are three parallel hedges of yew, the two outermost forming the "Dark Arbour," a tunnel of yew a hundred yards in length, only broken near its lower end, where a small fountain marks the crossing of a broad path.

All the lower portion of the garden is surrounded by a dense grove of trees, in which other tall Scotch Firs stand out conspicuously. Its most extensive area is on the right side of the Great Water, where several grassy paths, bounded by clipped hedges of yew and lime, radiate from a large circular space where there is a wide, round basin and fountain-jet. Looking along one of the pleasant green ways, other jets are seen springing from further fountains where more paths cross. The ends of some of the walks are finished with alcoves or arbours. One of them, that runs diagonally from the right-hand side of the large pool, crosses the great wood fountain, and passing on some distance further ends at a magnificent lead urn on a massive pedestal. This is also the terminal point of view of another of the longest of the green paths.

The water that supplies the pools and fountains comes from a wild pond, the home of many wild-fowl, that is on a higher level, outside the grounds and beyond one of the roads that bounds them. A stream from the pond meanders through the wooded ground, and is conducted by a culvert to the large pool; the overflow passing out on the opposite side in the same way.

Important in the garden's decoration are the unusual number of lead statues and other accessories, of excellent design. The upper lawn has two kneeling figures of negro or Indian type, bearing on their heads, partly supported by their hands, circular tables with moulded edges that carry an urn-finial. The central ornament of the next level is the Flying Mercury, after John of Bologna. Referring to this example, Messrs. Blomfield and Inigo Thomas tell us in "The Formal Garden in England" that "lead statues very easily lose their centre of gravity." This is exemplified by the Mercury at Melbourne, which has already come over to a degree which makes its evident want of balance distressing to the eye of the beholder, and forebodes its eventual downfall.

Lead as a material for such use in gardens is much more suitable to the English climate than marble. It acquires a beautiful silvery colouring with age, whereas marble becomes disfigured with blackish weather-streaks. During the eighteenth century the art of lead casting came to great perfection in England. Some good models came from Italy; the original of the kneeling slave at Melbourne is considered to have come from there. Others were brought from France. The inspiration, if not the actual designs or moulds, of the many charming figures of *amorini* in these

gardens must have been purely French. The pictures show how they were used. They stand on pedestals at several of the points of departure of the green glades. In fountain basins they form jets; the little figure appearing to blow the water through a conch-shell. They are also shown, sometimes singly, sometimes in pairs, disputing, wrestling or carrying a cornucopia of flowers. One little fellow, alone on his pedestal, is whittling his bow with a tool like a wheelwright's draw-knife. All are charming and graceful. They are probably more beautiful now than of old, when they were painted and sanded to look like stone.

MELBOURNE: AMORINI
FROM THE PICTURE IN THE POSSESSION OF
MR. J. W. FORD

There were several lead foundries in London early in the eighteenth century for the making of these garden ornaments. The foremost was that of John Van Nost. Mr. Lethaby in his book on Leadwork tells us that this Dutch sculptor came to England with King William III.; that his business was taken in 1739 by Mr. John Cheere, who served his time with his brother, Sir H. Cheere, who made several of the Abbey monuments. The kneeling slave, bearing either a vase, as at Melbourne, or a sundial as in the Temple Gardens in London, and in other pleasure grounds in different parts of the country, was apparently a favourite subject. The figure, not always from the same mould in the various examples, but always showing good design, was evidently of Italian origin. Towards the end of the century, designs for lead figures became much debased, and such subjects as people sitting round a table, painted like life, could not possibly have served any decorative purpose. The natural colour of lead is so good that no painting can improve it. In Tudor days it was often gilt, a much more permissible treatment.

In the old days there was probably a parterre at Melbourne, now no longer existing. The figures of kneeling slaves were possibly the centre ornaments of its two divisions, on what is now the upper lawn. This portion of the garden is rather liberally, and perhaps somewhat injudiciously, planted with a mixture of conifers, put in probably thirty to forty years ago, when the remains of good old garden designs were not so reverently treated, nor their value so well understood, as now. Some of this planting has even strayed to the banks of the Great Water. The pleasure ground of Melbourne is a precious relic of the past, and, even though the ill effects of

the modern planting of various conifers may be less generally conspicuous there than it is in many places, yet it is distinctly an intrusion. The tall trees inclosed by massive yew hedges, the pools and fountains, the statues and other sculptured ornaments, all recall, with their special character of garden treatment, the times and incidents that Watteau loved to paint. Such a picture as his *Bosquet de Bacchus*, so well known by the engraving, with its gaily-dressed groups of young men and maidens seated in the grassy shade and making the music of their lutes and voices accompany that of the fountains' waters, might have been painted at Melbourne. For here are the same wide, green-walled alleys, the pools, the fountains and the ornamental details of the great gardens of courtly France of two hundred years ago acclimatised on English soil; not in the dreary vastness of Versailles, but tamed to our climate's needs and on a scale attuned to the more moderate dimensions of a reasonable human dwelling.

BERKELEY CASTLE

THIS venerable pile, one of the oldest continuously-inhabited houses in England, stands upon a knoll of rising ground at the southern end of the tract of rich alluvial land known as the Vale of Berkeley, that stretches away for ten miles or more north-eastward in the direction of Gloucester. Within two miles to the west is the Severn, already a mile across and rapidly widening to its estuary. On the side of the higher ground the town creeps up to the shelter of the Castle and the grand old church, on the lower is a level stretch of water-meadow.

Seen from the meadows some half-mile away it looks like some great fortress roughly hewn out of natural rock. Nature would seem to have taken back to herself the masses of stone reared by man seven and a half centuries ago.

The giant walls and mighty buttresses look as if they had been carved by wind and weather out of some solid rock-mass, rather than as if wrought by human handiwork. But when, in the middle of the twelfth century, in the earliest days of the reign of Henry Plantagenet, the castle was built by Robert, son of Harding, he built it with outer walls ten to fifteen feet thick, without definite plan as it would seem, but, as the work went on, suiting the building to the shape of the hillock and to the existing demands of defensive warfare.

When the day is coming to its close, and the light becomes a little dim, and thin mist-films rise level from the meadows, it might be an enchanted castle; for in some tricks of evening light it cheats the eye into the semblance of something ethereal—sublimate—without substance—as if it

were some passing mirage, built up for the moment of towering masses of pearly vapour.

So does an ancient building come back into sympathy with earth and cloud. Its stones are carved and fretted by the wind and rain of centuries; tiny mosses have grown in their cavities; the decay of these has formed mould which has spread into every joint and fissure. Here grasses and many kinds of wild plants have found a home, until, viewed from near at hand, the mighty walls and their sustaining buttresses are seen to be shaggy with vegetation.

These immense buttresses on the meadow side come down to a walled terrace; their foundations doubtless far below the visible base. The terrace level is some twelve feet above the grassy space below. The grass then slopes easily away for a distance of a few hundred feet to the alluvial flat of the actual meadow-land.

Large fig-trees grow at the foot of the wall, rising a few feet above the parapet of the terrace, from which the fruit is conveniently gathered.

It is in the deep, well-sheltered bays between the feet of the giant buttresses that the most interesting of the modern flower gardening at Berkeley is done.

White Lilies grow like weeds in the rich red loam, and there are fine groups of many of the best hardy plants and shrubby things, gathered together and well placed by the late Georgina Lady Fitzhardinge, a true lover of good flowers and a woman of sound instinct and well-balanced taste respecting things beautiful both indoors and out.

The chief relic of the older gardening at Berkeley is the remains of the yew hedge that inclosed the bowling-green on three sides; the fourth side having for its boundary the high retaining wall that supports the entrance road beyond the outer gate. The yews, still clipped into bold rounded forms, may have formed a trim hedge in Tudor days, and the level space of turf, which is reached from the terrace by a flight of downward steps that passes under an arch of the old yews, lies cool and sheltered from the westering sun by the stout bulwark of their ancient shade.

The yew arch in the picture shows where the terrace level descends to the bowling-green. The great buttresses of the main castle wall are behind the spectator. A bowery Clematis is in full bloom over the steps

THE LOWER TERRACE, BERKELEY CASTLE
FROM THE PICTURE IN THE POSSESSION OF
MR. ALBERT WRIGHT

43

to the shorter terrace above, and near it, on the lower level, is one of the great pear-trees that have been trained upon the wall, and that, with others on the keep above, brighten up the grim old building in spring-time. *Campanula pyramidalis* has been sown in chinks on the inner side of the low parapet, and the picture shows how handsomely they have grown, supported only by the slight nutriment they could find among the stones. But, like so many of the Bell-flowers, it delights in growing between the stones of a wall. It should be remembered how well this fine plant will succeed in such a place, as well as for general garden use. It is so commonly grown as a pot-plant for autumn indoor decoration that its other uses would seem to be generally overlooked.

SUMMER FLOWERS

THE end of June and beginning of July—when the days are hot and long, and the earth is warm, and our summer flowers are in fullest mass and beauty—what a time of gladness it is, and of that full and thankful delight that is the sure reward for the labour and careful thoughtfulness of the last autumn and winter, and of the present earlier year!

The gardens where this reward comes in fullest measure are perhaps those modest ones of small compass where the owner is the only gardener, at any rate as far as the flowering plants are concerned; where he thinks out good schemes of plant companionship; of suitable masses of form and stature; of lovely colour-combination; where, after the day's work, comes the leisurely stroll, when every flower greets and is greeted as a close friend, and all make willing offering of what they have of scent and loveliness in grateful return for the past loving labour.

This is the high tide time of the summer flowers. It may be a week or two earlier or later according to the district, for our small islands have climatic diversities such as can only be matched within the greater part of the whole area of middle Europe, though inclining to a temperate average. For the Myrtle of the Mediterranean is quite hardy in the South and South-West, and Ivy and Gorse, neither of which is hardy in North and Middle Germany, are, with but few exceptions, at home everywhere. Given, therefore, a moderately good soil, fair shelter and a true love of flowers, there will be such goodly masses as those shown in the pictures.

Advisedly is the word "true" lover of flowers used, for it is now fashionable to like flowers, and much of it is pretence only. The test is to ascertain whether the person professing devotion to a garden works in

ORANGE LILIES AND LARKSPUR
FROM THE PICTURE IN THE POSSESSION OF
MR. GEORGE C. BOMPAS

46

it personally, or in any way likes it well enough to take a great deal of trouble about it. To those who know, the garden speaks of itself, for it clearly reflects individual thought and influence; and it is in these lesser gardens that, with rare and happy exceptions, the watchful care and happy invention of the beneficent individuality stamps itself upon the place.

There is nothing more interesting to one of these ardent and honest workers than to see the garden of another. Plants that had hitherto been neglected or overlooked are seen used in ways that had never been thought of, and here will be found new combinations of colour that had never been attempted, and methods of use and treatment differing in some manner to those that had been seen before.

There is nothing like the true gardening for training the eye and mind to the habit of close observation; that precious acquirement that invests every country object both within and without the garden's bounds with a living interest, and that insensibly builds up that bulk of mentally noted incident or circumstance that, taken in and garnered by that wonderful storehouse the brain, seems there to sort itself, to distribute, to arrange, to classify, to reduce into order, in such a way as to increase the knowledge of something of which there was at first only a mental glimpse; so to build up in orderly structure a well-founded knowledge of many of those things of every-day out-door life that adds so greatly to its present enjoyment and later usefulness.

So it comes about that some of us gardeners, searching for ways of best displaying our flowers, have observed that whereas it is best, as a general rule, to mass the warm colours (reds and yellows) rather together, so it is

best to treat the blues with contrasts, either of direct complementary colour, or at any rate with some kind of yellow, or with clear white. So that whereas it would be less pleasing to put scarlet flowers directly against bright blue, and whereas flowers of purple colouring can be otherwise much more suitably treated, the juxtaposition of the splendid blues of the perennial Larkspurs with the rich colour of the orange Herring Lily (*Lilium croceum*) is a bold and grand assortment of colour of the most satisfactory effect.

This fine Lily is one of those easiest to grow in most gardens. The true flower-lovers, as defined above, take the trouble to find out which are the Lilies that will suit their particular grounds; for it is generally understood that the soil and conditions of any one garden are not likely to suit a large number of different kinds of these delightful plants. Four or five successful kinds are about the average, and the owner is lucky if the superb White Lily is among them. But Lilies are so beautiful, so full of character, so important among other flowers or in places almost by themselves, that, when it is known which are the right ones to grow, those kinds should be well and rather largely used.

The garden in which these fine groups were painted has a good loamy soil, such as, with good gardening, grows most hardy flowers well, and therefore the grand White Lily also thrives. A few of the Lilies like peat, such as the great Auratum, and the two lovely pink ones, Krameri and Rubellum. But the garden of strong loam should never be without the White Lily, the Orange Lily, and the Tiger Lily, an autumn flower that seems to accommodate itself to any soil. The Orange Lilies are grandly

grown by the Dutch nurserymen in many varieties, under the names *bulbiferum*, *croceum*, and *davuricum*, and their price is so moderate that it is no extravagance to buy them in fair quantity.

Flowers of pure scarlet colour are so little common among hardy perennials that it seems a pity that the brilliant *Lilium chalcedonicum* of Greece, Palestine, and Asia Minor, and its ally *L. pomponium*, the Scarlet Martagon of Northern Italy, should be so seldom seen in gardens. They are some of the most easily grown, and are not dear to buy. Another Lily that should not be forgotten and is easy to grow in strong soils is the old Purple Martagon; not a bright-coloured flower, but so old a plant of English gardens that in some places it has escaped into the woods. The white variety is very beautiful, the colour an ivory white, and the flower of a waxy texture. They are the Imperial Martagon, or Great Mountain Lily of the old writers; the scarlet *pomponium*, of the same shaped flower, was their Martagon Pompony. The name "pompony," no doubt, came from the tightly rolled-back petals giving the flower something of the look of the flattened melons of the Cantaloupe kind, with their deep longitudinal furrows; the old name of these being "Pompion." Another name for this Lily was the Red Martagon of Constantinople. It is so named by that charming old writer Parkinson, who gives evidence of its

WHITE LILIES AND YELLOW MONKSHOOD
FROM THE PICTURE IN THE POSSESSION OF
Mr. Herbert D. Turner

popularity and former frequency in gardens in these words: "The Red Martagon of Constantinople is become so common everywhere, and so well known to all lovers of these delights, that I shal seem unto them to lose time, to bestow many lines upon it; yet because it is so fair a flower, and was at the first so highly esteemed, it deserveth his place and commendations, howsoever increasing the plenty hath not made it dainty." One more of the Lilies, indispensable for loveliness, should be grown wherever it is found possible. This is the Nankeen Lily (*L. testaceum*). It is a flower as mysterious as it is beautiful. It is not found wild, and is considered to be a hybrid between the White Lily and the Scarlet Martagon. Whether it occurred naturally, or whether it was the deliberate work of some unknown benefactor to horticulture, will now never be known; we can only be thankful that by some happy agency we have this Lily of mixed parentage, one of the most beautiful in cultivation. The name Nankeen Lily nearly, but not exactly, describes its colour, for a suspicion of pinkish warmth is added to the tender buff-colour usually so named.

Many other Lilies may be grown in different gardens, but the tenderer kinds from Eastern Asia are not for the hardy flower-border, and the vigorous American species have not yet been with us long enough to be familiar as flowers of old English gardens.

A July garden would not show its true character without some masses of the stately blue perennial Larkspurs. No garden plant has been more widely cultivated within the last fifty years, and our nurserymen have produced a large range of beautiful varieties. They have, perhaps, gone a

little too far in some directions. The desire to produce something that can be called a novelty often makes growers forget that what is wanted is the thing that is most beautiful, rather than something merely exceptionally abnormal, to be gaped at in wonderment for perhaps one season, and above all for the purpose of being blazoned forth in the trade list. The true points to look for in these grand flowers are pure colour, whether light, medium or dark, fine stature and a well-filled but not overcrowded spike. There are some pretty double flowers, where the individual bloom loses its normal shape and becomes flattened, but the single is the truer form. They are so easily raised from seed that good varieties may be grown at home, when, if space may be allowed for a line of seedlings in the trial-ground, it is pleasant to watch what they will bring forth. Such a good old kind as the one named "Cantab" is a capital seed-bearer, and will give many handsome plants. They must be carefully observed at flowering time, and any of poor or weedy habit in their bloom thrown away. Some will probably have interrupted spikes, that is to say, the spike will have some flowers below and then a bare interval, with more flowers above. This is a fault that should not be tolerated.

The Monkshoods (*Aconitum*) are related to the Larkspurs (*Delphinium*); indeed, it is a common thing to hear them confused and the name of one used for the other. It is easy to understand how this may be, for the leaves are much alike in shape, and both genera bear hooded flowers on tall spikes, mostly of blue and purple colours. For ordinary garden knowledge it may be remembered that Monkshood has a smooth leaf and that the colour is a purplish blue, the bluest of those commonly in cultivation

being the late-flowering *Aconitum japonicum*, and that the true pure blues are those of the perennial Larkspurs, whose leaves are downy.

The great Delphiniums love a strong, rich loamy soil, rather damp than dry, and plenty of nourishment.

There is a handsome Monkshood with pale yellow flowers that is well used in the garden of the White Lilies, and most happily in their near companionship. It is *Aconitum Lycoctonum*; a plant of Austria and the Tyrol. The widely-branched racemes of pale luminous bloom are thrown out in a graceful manner, in pleasant contrast with the equally graceful but quite different upright carriage of the White Lily. The handsome dark green polished leaves of this fine Aconite are also of much value; persisting after the bloom is over till quite into the late autumn.

Many of the charming members of the Bell-flower family are fine things in the flower-border. The best of all for general use is perhaps the well-known *Campanula persicifolia*, with its slender upright stems and its numbers of pretty bells, both blue and white. There are double

PURPLE CAMPANULA
FROM THE PICTURE IN THE POSSESSION OF
MISS BEATRICE HALL

kinds, but the doubling, though in some cases it makes a good enough flower, changes the true character so much that it is a Bell-flower no longer; and we think that a Bell-flower should be a bell, and should hang and swing, and not be made into a flattened flower set rather tightly on an ungraceful, thickened stem.

Another beautiful Campanula is *C. latifolia*, especially the white-flowered form. It is not only a first-rate flower, but it gives that pleasant impression of wholesome prosperity that is so good to see. The tall, pointed spike of large milk-white bells is of fine form, and the distinctly-toothed leaves are in themselves handsome. Like all the Bell-flowers, the bloom is cut into six divisions—"lobes of the corolla," botanists call them. Each division is sharply pointed and recurved or rolled back after the manner of many of the Lilies. This fine Campanula is not only a good plant for the flower-border, but also for half-shady places in quiet nooks where the garden joins woodland, in the case of those fortunate gardens that have such a desirable frontier-land; the sort of place where the instinct of the best kind of gardener will prompt him to plant, or rather to sow, the white Foxglove, and to plant the white French Willow (*Epilobium*).

Nothing is more commonly seen in gardens than wide-spread neglected patches of *Campanula grandis*. The picture shows it better grown. It spreads quickly and in many gardens flowers only sparingly, because the tufts should have been oftener divided. It is perhaps the most commonly grown of all, and though, as the picture shows, it can be more worthily used than is ordinarily done, it is by no means so pretty a plant as others of its family.

In good soils in our southern counties the tall and beautiful Chimney Campanula (*C. pyramidalis*), commonly grown in pots for the conservatory, should be largely used in the borders; it also loves a place in a wall joint. It is a plant that we are so used to see in a pot that we are apt to forget its great merit in the open ground.

Of the smaller Bell-flowers, *C. carpatica*, both blue and white, is one of the very best of garden plants; delightful from the moment when the first tuft of leaves comes out of the ground in spring till its full blooming time in middle summer. No plant is better for the front edge of a border, especially where the edge is of stone; though it is just tall enough to show up well over a stout box-edging.

The biennial Canterbury Bells are well known and in every garden. Their only disadvantage is that they flower in the early summer and then have to be cleared away, leaving gaps that may be difficult to fill. The careful gardener, foreseeing this, arranges so that their near neighbours in the border shall be such as can be led or trained over to take their places. It should not be forgotten that the Canterbury Bell is an admirable rock or wall plant, where the size of a rock-wall admits of anything so large. The wild plant from which it came has its home in rocky clefts in Southern Italy.

ROCKINGHAM

IN large gardens where ample space permits, and even in those of narrow limits, nothing is more desirable than that there should be some places, or one at least, of quiet greenery alone, without any flowers whatever. In no other way can the brilliancy of flowers be so keenly enjoyed as by pacing for a time in some cool green alley and then passing on to the flowery places. It is partly the unconscious working out of an optical law, the explanation of which in every-day language is that the eye, being, as it were, saturated with the green colour, is the more ready to receive the others, especially the reds.

Even in quite a small garden it is often possible to arrange something of the sort. In the case of a place that has just one double flower-border and a seat or arbour at the end, it would be easy to do by stopping the borders some ten feet away from the seat with hedges of yew or hornbeam, and putting other seats to right and left; the whole space being turfed.

The seat was put at the end in order to give the whole view of the border while resting; but, after walking leisurely along the flowers and surveying their effect from all points, a few minutes' rest on one of the screened side seats would give repose to the eye and brain as well as to the whole body, and afford a much better preparation for a further enjoyment of the flowers.

It was probably some such consideration that influenced the designers of the many old gardens of England, where yew, the grand walling tree, was so freely used. The first and obvious use was as a protection from wind and a screen for privacy, then as a beautiful background, and lastly perhaps

for resting and refreshing the eye, and giving it renewed appetite between its feasts of brilliant colouring and complex design. These green yew-bordered alleys occur without end in the old gardens. They were not always bowling-greens, though now often so called, but rather secluded ambulatories; places either for solitary meditation and refreshment of mind, or where friends would meet in pleasant converse, or statesmen hold their discourse on weightier matters. Such a place of cool green retreat is this straight alley of ancient yews. Almost better it might have been if the path were green and grassy too—Nature herself seems to have thought so, for she greens the gravel with mossy growths. Perhaps this mossiness afflicts the gardener's heart—let him take comfort in knowing how much it consoles the artist. Though a garden is for the most part the better for being kept trim, there are exceptional cases such as this, where to a certain degree it is well to let natural influences have their way. It is a matter respecting which it is difficult to lay down a law; it is just one for nice judgment. Had the path been freshly scratched up and rolled, and the verges trimmed to a perfectly true line, it would not have commended itself to the artist as a subject for a picture, but, as it is, it is just right. The mossy path is in true relation for colour to the trees and grassy edges, and the degree of infraction of the canons of orderliness stops short of an appearance of actual neglect.

Among the interesting features of the grounds at Rockingham is a rose-garden, circular in form, bounded and protected by a yew hedge. Four archways at equal points, cut in the hedge, with straight paths, lead to a concentric path within which is a large round bed, with poles and swinging

garlands of free-growing Roses. The outer quarters have smaller beds, some concentric, some parallel with the straight paths. The space is large enough to give ample light and air to the Roses, while the yew hedge affords that comforting shelter from boisterous winds that all good Roses love.

Close to the house a flight of steps leads to a flower garden on the higher level. A sundial on steps stands in the midmost space, with beds and clumps of bright flowers around. There is other good gardening at Rockingham, and a curious "mount"; not of the usual circular

THE YEW ALLEY, ROCKINGHAM
FROM THE PICTURE IN THE POSSESSION OF
MISS WILLMOTT

shape, but in straight terraces. But it is these grand old hedges of yew that seem to cling most closely to the fabric and sentiment of the ancient building—half house, half castle, whose windows have looked upon them for hundreds of years, and whose inmates have ever paced within their venerable shade.

BRYMPTON

BRYMPTON D'EVERCY in Somersetshire—not far from Montacute, the residence of the Hon. Sir Spencer Ponsonby-Fane—is a house of mixed architectural character of great interest. A large portion of the earlier Tudor building now shows as the western (entrance) front, while, facing southward, is the handsome façade of classical design, said to be the work of Inigo Jones, but more probably that of a later pupil. The balustraded wall flanking the entrance gates—the subject of the picture—appears to be of the time of this important addition, for it is better in design than the balustrade of the terrace, which was built in the nineteenth century.

But the terrace is of fine effect, with the great flight of steps midway in its length that lead down to a wide unspoilt lawn. This again passes to the fish-pond, then to parkland with undulating country beyond.

The treatment of the ground is admirable. Fifty years ago the lawn would probably have been cut up into flower-beds, a frivolity forbidden by the dignified front.

Gardening is always difficult, often best let alone, in many such cases. When the architecture, especially architecture of the classical type, is good and pure, it admits of no intrusion of other forms upon its surfaces. It is complete in itself, and the gardener's additions become meddling encroachments. When any planting is allowable against houses of this type—as in cases where they are less pure in style and have larger wall-spaces—it should be of something of bold leafage, or large aspect of one simple character; the strong-growing *Magnolia grandiflora* as an upright example, and *Wistaria* as one of horizontal

THE GATEWAY, BRYMPTON
FROM THE PICTURE IN THE POSSESSION OF
MR. EDWIN CLEPHAN

growth. There is some planting between the lower windows at Brympton, but it is doubtful whether it would not have been better omitted. It is a place more suitable (if on this front any gardening is desirable) for the standing of Bays or some such trees, in tubs or boxes on the terrace.

There is sometimes a flower-border at the base of such a house; where this occurs it is a common thing to see it left bare in winter and in the early year dotted with bulbous plants and spring flowers; to be followed in summer with bedding-plants. No such things look well or at all in place directly against a building. The transition from the permanent structure to the transient vegetation is too abrupt. At least the planting should be of something more enduring and of a shrubby character, and mostly evergreen. Such plants as *Berberis Aquifolium*, Savin, Rosemary and Laurustinus would seem to be the most suitable, with the large, persistent foliage of the Megaseas as undergrowth, Pyrus japonica for early bloom, and perhaps some China Roses among the Rosemary.

But happily this house has been treated as to its environment with the wisest restraint. No showy or pretentious gardening intrudes itself upon the great charm of the place, which is that of quiet seclusion in a beautiful but little-known part of the county. The place lies among fields—just the House, the Church and the Rectory. There is no village or public road. The house is approached by a long green forecourt inclosed by walls. Between this and the kitchen garden is the quiet, low, stone-roofed church, in a churchyard that occupies such another parallelogram as the forecourt. The pathway to the church passes across the forecourt into the restful churchyard with its moss-grown tombs and bushes of old-

fashioned Roses, and the grassy mounds that mark the last resting-place of generations of long-forgotten country folk.

The church has a bell-cote built upon the gable of its western wall of remarkable and very happy form, stone-roofed like the rest. Among the graves stands the base—three circular steps and a square plinth—of what was once an ancient stone cross. The church seems to lie within the intimate protection of the house, adding by its presence to the general impression of repose and peaceful dignity.

The picture shows the walled and balustraded entrance, probably contemporary with the classical façade, wrought of the local Ham Hill stone; a capital freestone for the working of architectural enrichment. It is of a warm yellowish-brown colour; but grey and yellow lichens and brown mosses have painted the surface after their own wayward but always beautiful manner. A light cloud of *Clematis Flammula* peeps over the bushes through the balusters. Stonework so good as this can just bear such a degree of clothing with graceful flowery growth; no doubt it is watched and not allowed to hide too much with an excess of overgrowth. Where garden architecture is beautiful in proportion and detail it is not treating it fairly to smother it with vegetation. How many beautiful old buildings are buried in Ivy or desecrated by the unchecked invasion of Veitch's Virginia Creeper!

BALCASKIE

EQUIDISTANT from Pittenweem and St. Monan's, in Fifeshire, and a mile from the sea, stands Balcaskie, the beautiful home of Sir Ralph Anstruther. The park is entered from the north by a fine gateway with stone piers bearing "jewelled" balls, dating from the later middle of the seventeenth century. The entrance road is joined by two others from east and west, all passing through a park of delightful character. The road leads straight through a grassy forecourt walled on the three outer sides by yew hedges, and reaches the door by a gravelled half-circle formed by the projection on either side, of the curved walls of the offices and stables. The house, of the middle seventeenth century, though just too late to have been built as a fortress, retains much of the character of the older Scottish castles, but adds to it the increased comfort and commodiousness of its own time. There have been considerable later additions and alterations, but much of the old still remains, including some rooms with very interesting ceilings.

The main entrance on the north leads straight through to a door to the garden on the south. The garden occupies a space equal to about five times the length of the house-front. The ground falls steeply, something like fifty feet in all, and is boldly terraced into three levels. Looking southward from the door and across the garden, the eye passes down a great vista between trees in the park to the Firth of Forth, and across it to the Bass Rock, some twelve miles away and near the further shore.

The upper garden level, reached from the house by a double flight of descending steps, has a broad walk running the whole length, with an excellently modelled lead statue at each end; to the west an Apollo, a

singularly graceful figure, and to the east a female statue, possibly a Diana. The space in front of the house is divided into three portions; the two outer compartments having hedges of yew from four to five feet high. One of these incloses a bowling-green, the other a lawn with some beds. The middle turfed space has a sundial and beds of flowers. Here is also the remaining one of what was formerly a pair of fine cedars, placed symmetrically to right and left. Adjoining the house and next to the end of the broad walk where stands the Apollo, is the rose-garden, which, with this graceful statue, forms the subject of the picture. The rose-garden is of beds cut in the grass, containing not Roses only but also other bright garden flowers. A female statue of more modern work stands in the centre.

The great terrace wall, eighteen feet high, that forms the retaining wall of the upper portion of the garden, rises towards both ends to its full height as a wall, but the middle space is lightened by being treated with a handsome balustrade. At the extreme ends flights of steps lead down to the next, the middle level. The first long flight reaches a wide stone landing, the lower, shorter flight turning inwards at a right angle. Great buttresses, projecting forward eight feet at the ground-line, add much to the dignity and beauty of the wall. They are roofed with stone, and each one carries the bust of a Roman emperor. From the steps on each side come broad gravelled walks, leading by one step down to a slightly sunk rectangular lawn, which occupies the middle space. On each side of the paths are groups of flower-beds on a long axial line that is parallel with the wall. They have a broad turf verge and a nearly equal space of gravel next

to their box-edges. Piers and other important points have stone balls or flower-vases. Stone seats stand upon the landings above the lowest flights or steps, against the walls which bound the garden to right and left. Beyond these boundaries are tall trees, their protecting masses giving exactly that comforting screen that the eye and mind desire, and forming the best possible background to the structure and garnishing of the beautiful garden.

It is one of the best and most satisfying gardens in the British Isles;

THE APOLLO, BALCASKIE
FROM THE PICTURE IN THE POSSESSION OF
Miss Bompas

Italian in feeling, and yet happily wedding with the Scottish mansion of two and a half centuries ago, and forming, with the house and park-land, one of the most perfect examples of a country gentleman's place. All of it is pleasant and beautiful, home-like and humanly sympathetic; the size is moderate—there is nothing oppressively grand.

More than once already in these pages attention has been drawn to the danger of letting good stone-work become overgrown with rank creepers. At Balcaskie this is evidently carefully regulated. The wall-spaces between the great buttresses, and the buttresses themselves, are sufficiently clothed but never smothered with the wall-loving and climbing plants. The right relation of masonry and vegetation is carefully observed; each graces and dignifies the other; the balance is perfect.

The lowest level is given to the kitchen garden. It is not put out of the way, but forms part of the whole scheme. It is reached by a single flight of handsome balustraded stone steps.

Balcaskie occurs as a place-name early in the thirteenth century. From 1350 to 1615 it was owned by a family named Strang, afterwards by the Moncrieffs, till 1665. It is not known whether any portion of the present house and garden belonged to these earlier dates, but it is probable that the designer of both was Sir William Bruce, one of the best architects of the time of Charles II., and an owner of Balcaskie for twenty years.

CRATHES CASTLE

CRATHES CASTLE in Kincardineshire presents one of the finest examples of Scottish architecture of the sixteenth century. It is the seat of Sir Robert Burnett of Leys, the eleventh baronet and descendant of the founder.

Profoundly impressive are these great northern buildings, rising straight and tall out of the very earth. As to their lower walls, they are grim, forbidding, almost fiercely repellent. There is an aspect of something like ruthless cruelty in the very way they come out of the ground, without base or plinth or any such amenity—built in the old barbarous days of frequent raiding and fighting, and constant need of protection from marauders; when a man's house must needs be a strong place of defence.

This is the first impression. But the eye travelling upward sees the frowning wall blossom out above into what has the semblance of a fairy palace. It is like a straight, tall, rough-barked tree crowned with fairest bloom and tenderest foliage. Turrets both round and square, as if in obedience to the commanding wave of a magician's wand, spring out of the angles of the building and hang with marvellous grace of poise over the abyss. There seems to be no actual plan, and yet there is perfect harmony; the whole beautiful mass appears as if it had come into being in some one far-away, wonderful, magical night! It is a sight full of glamour and romantic impression—grim fortalice below, ethereal fantasy aloft. Rough and rugged is the rock-like wall, standing dark and dim in the evening gloom; intangible, opalescent are the mystic forms above, in the tender warmth of the afterglow; cloud-coloured, faintly rosy, with shadows pearly-blue.

THE YEW WALK, CRATHES

FROM THE PICTURE IN THE POSSESSION OF

MR. CHARLES P. ROWLEY

Direct descendants of the old Norman keep, these Scottish castles, for the most part, retain the four-sided tower, as to the main portion of the structure. The walls need no buttresses, for they are of immense thickness, and the vaulted masonry, usually of the simple barrel form, that carries the floors of, at any rate, the lower stories, ties the whole structure together. The angle turrets carried on bold corbels that are so conspicuous a feature of these northern castles, broke away from the Norman forms and became a distinct character of the Scottish work. They were a helpful addition to the means of defence, and, as long as they were built for use, added much to the beauty and dignity of the structure. The only detail that shows a tendency to debasement in Crathes is the quantity of useless cannon-shaped gargoyles, put for ornament only, in places where they could not possibly do their legitimate work of carrying off rain-water from the roof.

There could have been no pleasure garden in the old days; but now these ancient strongholds, mellowed by the centuries, seem grateful for the added beauty of good gardening. The grand yew hedges may be of the seventeenth century. They stand up solid and massive for ten feet or more, with roof-shaped tops, and then rise again at intervals into great blocks, bearing ornaments like circular steps crowned with a ball. The ornament is simpler, a low block and ball only, in the first picture, where they accentuate the arches that lead right and left into the two divisions of the flower garden. This plainer form is perhaps more suitable to this grand old place than the more elaborate, just because it is simpler and more dignified.

The flower garden, as it is to-day, is quite modern. The finest of the hardy flowers are well grown in bold groups. Luxuriant are the masses of Phlox and tall Pyrethrum, of towering Rudbeckia, of Bocconia, now in seed-pod but scarcely less handsome than when in bloom; of the bold yellow Tansy and Japan Anemones; all telling, by their size and vigour, of a strong loamy soil.

Many are the arches of cluster and other climbing Roses; at one point in the kitchen garden coming near enough together to make a tunnel-like effect.

Wonderful is the colouring and diversity of texture!—the bright flowers, the rich, dark velvet of the half-distant yews, the weather-worn granite and rough-cast of the great building.

If the flowers in the second and third pictures were in our southern counties the time would be the end of August or at latest the middle of September, but the seasons of the flowers in Scotland are much later, and these would be October borders.

The Castle stands upon a wide, level, grassy terrace, which is stopped on the north-eastern side by the parapet of a retaining wall, broken by a flight of steps down to the path that is bounded by the two hedges of ancient yews shown in the first picture. These hedges divide the flower garden into two equal parts on the lower level, for, from where the Castle stands, the ground falls to the south and east. On each side of the steps, just beneath the terrace wall, is a flower border. Immediately on entering the double wall of yew there is an opening to right and left—an arch cut in the living green—giving access to the two square gardens, in both of which a

path passes all round next the yews. There is also a flower border on two sides. The middle space is grass with flower beds; in the left-hand garden (coming from the Castle) are bold masses of herbaceous plants in beds grouped round a fountain; in the one on the right, for the most part, Roses and Lilies.

To the south-east, and occupying the space next beyond the rose garden and the end of the lawn adjoining the Castle, is the kitchen garden. The main walks have flower borders. Where the two cross paths intersect is a Mulberry tree with an encircling seat. The subjects of the second and third pictures are within the kitchen garden.

Many are the beautiful points of view from the kitchen garden, for there the grand yew hedges show beyond the flowers; then, towering aloft, comes the fairy castle, and then fine trees; for trees are all around, closely approaching the garden's boundaries.

The brilliancy of colour masses in these Scottish gardens is something remarkable. Whether it is attributable to soil or climate one cannot say; possibly the greater length of day, and therefore of daily sunshine, of these northern summers, may account for it. Of the great number of people who go North for the usual autumn shooting, those

CRATHES
FROM THE PICTURE IN THE POSSESSION OF
Mr. George C. Bompas

who love the summer flowers find their season doubled, for the kinds they have left waning in the South are not yet in bloom in the more northern latitude. The flowers of our July gardens, Delphiniums, Achilleas, Coreopsis, Eryngiums, Geums, Lupines, Scarlet Lychnis, Bergamot, early Phloxes, and many others, and the hosts of spring-sown annuals, are just in beauty. Sweet Peas are of astounding size and vigour. Strawberries are not yet over, and early Peas are coming in. The Gooseberry season, that had begun in the earliest days of August with the Early Sulphurs and had

76

been about ten days in progress in the Southern English gardens, is for a time interrupted, but resumes its course in September in the North, where this much-neglected fruit comes to unusual excellence. It is a hardy thing, and appears to thrive better north of the Border than elsewhere.

It is one of the wholesomest of fruits; its better sorts of truly delicious flavour. It is a pleasure, to one who knows its merits, to extol them. It is essentially a fruit for one who loves a garden, because, for some reason difficult to define, it is less enjoyable when brought to table in a dessert dish. It should be sought for in the garden ground and eaten direct from the bush. Perhaps many people are deterred by its spiny armature, and it is certain that, when, as is too often the case, the bushes are in crowded rows and have been allowed to grow to a large size, the berries are difficult to get at.

But the true amateur of this capital autumn fruit has them in espalier form, in a few short rows, with ample space—about six feet—between each row. The plants may be had ready trained in espalier shape, but it is almost as easy to train them from the usual bush form. The vigorous young growth that will spring out every year is cut away at the sides in middle summer; just a shoot or two of young wood being left, when the bushes have grown to a fair size, to train in, to take the place of older wood. The plants being restricted to the fewer branches that form the flat espalier, more strength is thrown into the ones that remain, so that the berries become larger; and, as plenty of light and sun can get to the fruits, even the best kinds are sweeter and better flavoured than when they are allowed to grow in dense bushes.

Then when the kinds are ripe how pleasant it is to take a low seat and sit at ease before each good sort in succession! The best and ripest fruits can be seen at a glance and picked without trouble, in pleasant contrast to the painful, prickly groping that goes on among the crowded bushes. No one would ever regret planting such excellent sorts as Red Champagne, Amber Yellow, Cheshire Lass, Jolly Painter, a large, well-flavoured and little-known berry, and Red Warrington, a trusty late kind. To these should be added two admirable Gooseberries lately brought out by Messrs. Veitch, namely, Langley Green and Langley Gage, both fine fruits of delicious flavour.

If such a little special fruit space were planted in these large Scottish gardens, and the merits of the kinds became known, the daily invitation of the hostess, "Let us go to the gooseberry garden," would be gladly welcomed, and guests would also find themselves, at various times of day, sauntering towards the gooseberry plot.

How grandly the scarlet Tropæolum (*T. speciosum*) grows in these northern gardens is well known; indeed, in many places it has become almost a pest. It is much more difficult to grow in the South, where it is often a failure; in any case, it insists on a northern or eastern exposure. Where it does best in gardens in the English counties is in deep, cool soil, thoroughly enriched. When well established, the running roots ramble in all directions, fresh growths appearing many feet away from the place where it was originally planted. It looks perhaps best when running up the face of a yew hedge, when the bright scarlet bloom, and leaves of clear-cut shape, are

seen to great advantage, and many of the free growths of the plant take the form of hanging garlands.

CRATHES: PHLOX
FROM THE PICTURE IN THE POSSESSION OF
Mrs. Croft

79

KELLIE CASTLE

KELLIE CASTLE in Fifeshire, very near Balcaskie, is another house of the finest type of old Scottish architecture. The basement is vaulted in solid masonry, the ground-floor rooms have a height of fourteen feet; the old hall, now the drawing-room, is nearly fifty feet long. A row of handsome stone dormers to an upper floor, light a set of bedrooms, which, as well as the main rooms below, have coved plaster ceilings of great beauty.

There is no certain record of the date of the oldest part of the castle. It is assigned to the fourteenth century, but may be older. The earliest actual date found upon the building is 1573, and it is considered that the mass of the castle, as we see it now, was completed by that date, though another portion bears the date 1606. It belonged of old to the Oliphants, a family that held it for two and a half centuries, when it passed by sale to an Erskine, who, early in the seventeenth century, became Earl of Kellie. In 1797, after the death of the seventh Earl, it was abandoned by the family and soon showed signs of deterioration from disuse. About thirty years later the Earldom of Kellie descended to the Earl of Mar, and the family seat being elsewhere, Kellie was allowed to go to ruin.

In 1878 the ruined place was taken, to its salvation, on a long lease by Mr. James Lorimer, whose widow is the present occupier. It has undergone the most careful and reverent reparation. The broken roofs have been made whole, the walls are again hung with tapestries, and the rooms furnished with what might have been the original appointments.

The castle stands at one corner of the old walled kitchen garden, a door in the north front opening directly into it. The garden has no architectural

features. There are walks with high box edgings and quantities of simple flowers. Everywhere is the delightful feeling that there is about such a place when it is treated with such knowledge and sympathy as have gone to the re-making of Kellie as a delightful human habitation. For two sons of the house are artists of the finest faculty—painter and architect—and they have done for this grand old place what boundless wealth, in less able hands, could not have accomplished.

Close to the house on its western side is a little glen, and in it a rookery. When strong winds blow in early spring the nests in the swaying tree-tops come almost within hand reach of the turret windows of the north-west tower.

How the flowers grow in these northern gardens! Here they must needs grow tall to be in scale with the high box edging. But Shirley Poppies, when they are autumn sown, will rise to four feet, and the grand new strains of tall Snapdragons will go five and even over six feet in height.

As the picture shows, this is just the garden for the larger plants—single Hollyhocks in big free groups, and double Hollyhocks too, if one can be sure of getting a good strain. For this is just the difficulty. The strains admired by the old-fashioned florist, with the individual flowers tight and round, are certainly not the best in the garden. The beautiful double garden Hollyhock has a wide outer frill like the corolla of the single flowers in the picture. Then the middle part, where the doubling comes, should not be too double. The waved and crumpled inner petals should be loosely enough arranged for the light to get in and play about, so that in some of them it is reflected, and in some transmitted. It is only in such

flowers that one can see how rich and bright it can be in the reds and roses, or how subtle and tender in the whites and sulphurs and pale pinks. Other flowers beautiful in such gardens are the taller growing of the Columbines, the feathery herbaceous Spiræas, such as *S. Aruncus*, that displays its handsome leaves, and waves its creamy plumes, on the banks of Alpine torrents, and its brethren the lovely pale pink *venusta*, the bright rosy *palmata* and the cream-white *Ulmaria*, the

KELLIE CASTLE
FROM THE PICTURE IN THE POSSESSION
Mr. Arthur H. Longman

garden form of the wild Meadow-sweet of our damp meadow-ditches. Then the tall Bocconia, with its important bluish leaves and feathery flower-beads, which shows in the picture in brownish seed-pod; and the Thalictrums, pale yellow and purple, and Canterbury Bells, and Lilies yellow and white, and the tall broad-leaved Bell-flowers.

All these should be in these good gardens, besides the many kinds of Scotch Briers, and big bushes of the old, almost forgotten garden Roses of a hundred years ago, many of which are no longer to be found, except now and then in these old gardens of Scotland. For here some gardens seem to have escaped that murderously overwhelming wave of fashion for tender bedding plants alone, that wrought such havoc throughout England during three decades of the last century.

Here, too, are Roses trained in various pretty simple ways. Our garden Roses come from so many different wild plants, from all over the temperate world, that there is hardly an end to the number of ways in which they can be used. Some of them, like the Scotch Briers, grow in close bushy masses; some have an upright habit; some like to rush up trees and over hedges; others again will trail along the ground and even run downhill. Some are tender and must have a warm wall; some will endure severe cold; some will flower all the summer; others at one season only. So it is that we find in various gardens, Roses grown in many different ways. In one as small bushes in beds, or budded on standards, in another as the covering of a pergola, or as fountain roses, throwing up many stems which arch over naturally. Some of the oldest garden Roses, such as *The Garland, Dundee Rambler* and *Bennett's Seedling* are the best for this kind of use.

The Himalayan *R. polyantha* will grow in this way into a huge bush, sometimes as much from thirty to forty feet in diameter, and many of the beautiful modern garden Roses that have *polyantha* for a near ancestor, will do well in the same way, though none of them attain so great a size. Roses grown like this take a form with, roughly speaking, a semi-circular outline, like an inverted basin. If they are wanted to take a shape higher in proportion they must be trained through or over some simple framework. This is called balloon-training. Some roses are grown in this fashion at Kellie, the framework being a central post from which hoops are hung one above the other. The Rose grows up inside the framework and hangs out all over. If this kind of training is to be on a larger scale, long half-hoops have their ends fixed in the ground, and pass across and across one another at a central point, where they are fixed to a strong post, thus forming ten or twelve ribs. Horizontal wires, like lines of latitude upon a globe, pass all round them at even intervals. Then Roses can be trained to any kind of trellis, either a plain one to make a wall of roses or a shaped one, whose form they will be guided to follow. Then again, there may be rose arches, single, double or grouped; or in a straight succession over a path; or alternate arches and garlands, a pretty plan where paths intersect; the four arches kept a little way back from the point of intersection, with garlands connecting them diagonally in plan. Then there are Roses, some of the same that serve for several of these kinds of free treatment, for making bowers and arbours.

And there are endless possibilities for the beautiful treatment of Rose gardens, though seldom does one see them well done. There are many

who think that a Rose garden must admit no other flowers but Roses. This may be desirable in some cases, but the present writer holds a more elastic view. Beds and clumps of Roses where no other flower is allowed, often look very bare at the edges, and might with advantage be under-planted with Pinks and Carnations, Pansies, London Pride, or even annuals. And any Lilies of white and pink colouring such as *candidum*, *longiflorum*, *Brownii*, *Krameri*, or *rubellum* suit them well, also many kinds of Clematis. The gardener may perhaps, object that the usual cultivation of Roses, the winter mulch and subsequent digging in and the frequent after-hoeing precludes the use of other plants; but all these rules may be relaxed if the Rose garden is on a fairly good rose soil. For the object is the showing of a space of garden ground made beautiful by garden Roses—not merely the production of a limited number of blooms of exhibition quality.

The way the bushes of garden Roses grow and bloom in close companionship with other strong-growing plants, at Kellie and in thousands of other gardens, shows how amicably they live with their near neighbours; and often by a happy accident, they tell us what plants will group beautifully with them.

The Roses that are best kept out of the Rose garden, are those delightful ones of the end of June; the Damasks, and the Provence, the sweet old Cabbage Rose of English gardens. These, and the Scotch Briers of earlier June, bloom for one short season only. Of late years the possibilities of beautiful Rose gardening have been largely increased by the raising of quantities of beautiful Roses of the Hybrid Tea class that bloom

throughout the summer, and that, with the coming of autumn, seem only to gain renewed life and strength.

HARDWICK

HARDWICK HALL in Derbyshire, one of the great houses of England, is, with others of its approximate contemporaries of the later half of the sixteenth century, such as Longleat, Wollaton, and Montacute, an example of what was at the time of its erection an entirely new aspect of the possibilities of domestic architecture.

The country had settled down into a peaceful state. A house was no longer a castle needing external defence. Hitherto the homes of England had been either fortresses, or had needed the protection of moats and walls. They had been poorly lighted; only the walls looking to an inner court, or to a small walled garden could have fair-sized openings. No spacious windows could look abroad upon open country, field or woodland. But by this time such restriction was a thing of the past, and we see in these great houses, and in Hardwick especially, immense window spaces in the outer walls. The architects of the time, John Thorpe, the Smithsons and others, ran riot with their great windows, as if revelling in their exemption from the older bonds. The new freedom was so tempting that they knew not how to restrain themselves, and it was only later, when it was found that the amount of lighting was overmuch for convenience, that the relation of degree of light to internal comfort came to be better understood and more reasonably adjusted.

The famous Countess of Shrewsbury (Bess of Hardwick), to whose initiative this great house owes its origin, set an imperishable memorial of her imperious arrogance upon the balustrading that crowns the square tower-like projections at the angles and ends of the building, where the

THE FORECOURT: HARDWICK
FROM THE PICTURE IN THE POSSESSION OF
Mr. Aston Webb

stone is wrought into lace-like fretwork of arabesque, whereof the chief features are her coronet and the initials of her name.

A spacious forecourt occupies the ground upon the western—the main entrance front. It stretches the whole length of the house, and projects as much forward; its outer sides being inclosed with a wall that bears in constant succession an ornament of a *fleur-de-lys* with tall pyramidal top, a detail imported direct from Italy, from the Renaissance gardens of earlier date. Such an ornament occurs at the Villa d'Este at Tivoli, crowning a

retaining wall. The entrance to the inclosed forecourt is by a handsome stone gateway. This gateway forms the background of the picture, which shows one of the well-planted flower borders that abound at Hardwick, and that strike that lightsome and cheerful note of human care and delight that is so welcome in this place whose scale is rather too large, and somewhat coldly forbidding, in relation to the more ordinary aspects of daily comfort.

Indeed—for all the good planting—the long wall-backed flower border facing south, whose wall is in part of its length that of the house itself, looks as if, in relation to the great building towering above it—its occupants were still too small, although they include flowering plants seven to nine feet high, such as Gyneriums and the larger herbaceous Spiræas. A well-directed effort has evidently been made to have the planting on a scale with the lordly building, but the items want to be larger still and the grouping yet bolder, to overcome the dwarfing effect of the towering structure. In such a place the Magnolias, both evergreen and deciduous, would have a fine effect, though possibly they would hardly thrive in the midland climate.

Within the forecourt, along the wall parallel to the house and furthest from it, this need is not so apparent. In the subject of the picture, the Honeysuckle, the magnificently grown purple Clematis upon the wall, the Mulleins, Bocconia and Japan Anemones, are in due proportion; the Tufted Pansies and Mignonette bringing their taller brethren happily down to the grassy verge. Approaching the pathway from the right, stretch some

of the long loose growths of one of the two large Cedars that are such prominent objects in the forecourt garden.

The main open spaces of this garden repeat in flower beds on grass the big E.S. of the self-asserting founder. It is not pretty gardening nor particularly dignified. No doubt it is only a modern acquiescence in the dominating tradition of the place. Even making allowance for, and retaining this sentiment, a better design might have been made, embodying these already too-often-repeated letters. Moreover, the servile copying of the lettering in its stone form only serves to illustrate the futility of reproducing a form of ornament designed for one material in another of totally different nature.

There is some excellent gardening in a long flower-border outside the forecourt wall. Here the size of the house is no longer oppressive, and it comes into proper scale a little way beyond the point where the broad green ways, bounded by noble hedges of ancient yews, swing into a wide circle as they cross, and show the bold niches cut in the rich green foliage where leaden statues are so effectively placed.

By the kindness of the owner, the Duke of Devonshire, Hardwick Hall, illustrating as it does a distinct form of architectural expression with much of historical interest, is open to the public.

MONTACUTE

MONTACUTE in Somersetshire, built towards the end of the sixteenth century by Sir Edward Phelips, is another of that surprising number of important houses built on a symmetrical plan that arose during the reign of Queen Elizabeth.

As the house was then, so we see it now; unaltered, and only mellowed by time. The gardens, too, are of the original design, including a considerable amount of architectural stonework.

The large entrance forecourt is inclosed by a high balustraded wall, with important and finely-designed garden houses on its outer angles. The length of the side walls is broken midway on each side by a small circular pillared pavilion with a boldly projecting entablature, crowned with an openwork canopy and a topmost ornament of two opposite and joining rings of stone.

The piers of the balustrade are surmounted by stone obelisks, and the large paved landing, forming a shallow court at the top of the flight of steps a hundred feet wide, that gives access to the house on this side has tall pillars that now carry lamps, though they appear to have been designed merely as a stately form of ornament.

The forecourt has a wide expanse of gravel with a large fountain basin in the middle. Next the wall there are flower-borders; then the wide gravelled path, and, following this, a broad strip of turf with Irish yews at regular intervals. The general severity of the planning is pleasantly relieved by the bright flower-border, the subject of the picture. To right and left are openings in the wall leading to other garden spaces. The one of these to

the left, just behind the spectator as in the picture, leads by an upward flight of steps to one side of a wide terrace walk, that encompasses on all four sides a large sunk garden of formal design. This garden runs the whole length of the forecourt and depth of the house, and has a width equal to some two-thirds of its length. A large middle fountain-basin, with shaped outline of angles and segments of circles, has a balustraded kerb with a stone obelisk on every pier. In the centre is a handsome tazza in which the water plays. Wide paths lead down flights of balustraded steps from all four sides to the gravelled area within which the fountain stands. The spaces between, and the banks rising to the level of the upper terraces, are of turf. Rows of Irish yews stand ranged on both levels. It is all extremely correct, stately—dare one say a trifle dull? Opposite the forecourt the garden is bounded by a good yew hedge protecting it from wind from the valley below. Midway in the length is an opening where a low wall and seats give a welcome outlook. The same yew hedge returns eastward to the south-east angle of the house; the garden's opposite boundary being a low wall with a sunk fence outside, giving a view into the park.

There is an entrance from the garden to the house on its southern side by a flight of balustraded steps, and niches with seats are on either side of the door.

Wonderful are these great stone houses of the early English Renaissance— wonderful in their bold grasp and sudden assertion of the new possibilities of domestic architecture! For it may be repeated that it was only of late that a man's house had ceased to be a place of defence, and that he might

venture to have windows looking abroad all round, and yet feel perfectly safe without even an inclosing moat.

In the present day it is somewhat difficult to account for the designer's attitude of mind when deciding on such a lavish employment of the obelisk-shaped finials. One can only regard it as the outcome of the taste or fashion of the day, when he borrowed straight from the Italians everything except their marvellous discernment. One accepts the many obelisks at Montacute as showing the reflection of Italian influence on the Tudor mind; to-day and new, they would be inadmissible. The modern mind, with the vast quantity of material at hand, and the easy access to all that has been said and done on the subject, should

MONTACUTE: SUNFLOWERS
FROM THE PICTURE IN THE POSSESSION OF MR. E. C. AUSTEN
LEIGH

accept nothing but the best and purest in this as in any other branch of fine art.

There is one other possible way of accounting for the prevalence of these all-pervading obelisks. The name of the place is taken from a conical wooded hill (*mons acutus*). The same play on a word, a favourite fashion of Elizabethan times, and a custom in heraldry from a remoter antiquity, is seen in the shield of the ancient Montacute family, where the three sharp peaks denote that the surname had the same origin. The connexion of this name with the acute peak or obelisk form would therefore the more readily commend itself to the Elizabethan mind.

The house has never gone into other hands, the present owner, Mr. W. R. Phelips, being the descendant of the founder.

RAMSCLIFFE

IT would seem to be a law that the purest and truest human pleasure in a garden is attained by means whose ratio is exactly inverse to the scale or degree of the garden's magnificence. The design, for instance, of a Versailles impresses one with a sense of ostentatious consciousness of magnitude; out of scale with living men and women; whose lives could only be adapted to it, as we know they were, by an existence full of artificial restraints and discomforts; the painful and arbitrarily imposed conditions of a tyrannical and galling etiquette.

So we think also of our greatest gardens, such as Chatsworth. It is visited by a large number of people who go to see it as a large expensive place to gape at, but surely not for the truest love of a garden. So it is with many a large place; the size and grandeur of the garden may suit the great house as a design; it may be imposing and costly, it may be beautifully kept, and yet it may lack all the qualities that are needed for simple pleasure and refreshment. It is not till we come to some old garden of moderate size that has always been cherished and has never been radically altered, that the true message of the garden can be received and read; and it is from thence downward in the scale of grandeur that we find those gardens that are the happiest and best of all for true delight and close companionship; the simple borders of hardy flowers, planted and tended with constant watchfulness and loving care by the owner's own hands.

Such a garden is this of Mr. Elgood's; in a midland county, and on a strong soil that throws up good hardy plants in vigorous luxuriance. Here grow the great Orange Lilies—the Herring Lilies of

RAMSCLIFFE: ORANGE LILIES & MONKSHOOD
FROM THE PICTURE IN THE POSSESSION OF Mr. C. E.
FREELING

the Dutch, because they bloom at the time of the herring harvest—six and seven feet high, and with them the Monkshood, with its tall spikes of hooded bloom. In poorer soils or with worse culture these fine flowers are of much lower growth, the Monkshood often only half the height, with its deeply-cut leaves yellowing before their time with the weakness of too-early maturity. The pleasure with which one sees this fine old garden flower is, however, always a little lessened by the knowledge of the dangerously poisonous nature of the whole plant, and especially of the root. It is the deadly Aconite of pharmacy. Another of the same family is

96

grouped with it; the yellow Aconite of the Austrian mountains, with branched heads of sulphur-coloured bloom and singularly handsome leaves—large, dark green, glistening and persistently enduring—for, long after the bloom is past, they are beautiful in the border.

How well an artist knows the value of grey-leaved plants, and their use in pictorial gardening in the way of giving colour-value by close companionship, to tender pinks and lilacs, and, above all, to whites! A patch of white bloom is often too hard and sudden and inharmonious to satisfy the trained eye, but led up to and softened and sweetened by masses of neighbouring tender grey it takes its proper place and comes to its right strength in the well-ordered scheme. Lavender, Lavender-cotton (*Santolina*), Catmint, Pinks and Carnations, and the Woolly Woundwort (*Stachys*) with some other plants of hoary foliage, do this good work. In this garden the Woundwort, there known by its old Midland name of "Our Saviour's Blanket," throws up its grey-pink heads of bloom from a thick carpet of rather large leaves, silvery soft with their thick coating of long white down. Here a groundwork of it leads to the group of white Peach-leaved Bell-flower on the right and to the tall white Gnaphalium, a plant of kindred woolliness, on the left, while the precious grey quality runs through the left-hand flower-group by means of the downy-coated pods of the earlier-blooming Lupins, purposely left among the later flowers for this and for their handsome form.

How finely the Orange Lilies tell against the background of the holly hedge, at the path-end cut into an arbour, may well be seen in the picture,

and how kindly and gracefully the Greengage Plum-tree bends over and plays its appointed part.

Such a flower border makes many a picture in the hands of a garden-artist. His knowledge of the plants, their colours, seasons, habits and stature, enables him to use them as he uses the colours on his palette.

How grandly the tall Delphiniums grow in this strong soil. A little of the colour has been lost owing to technical difficulties of reproduction, for the blue is purer and stronger in effect both in the original picture and in nature than is here shown. They are grouped, as blue flowers need, with contrasts of yellow and orange; with yellow Daisies and the feathery Meadow-rue (*Thalictrum*), and the tall yellow Aconite and nearly white Campanulas, woolly Stachys and purple Bell-flowers beyond. Only one small patch of brighter colour, the scarlet of Lychnis chalcedonica, is allowed here. On the other side is the loose-growing and always pictorial white Mallow (*Sidalcea candida*), taking some weeks to produce its crop of flowers that, like Foxgloves and most of the flowers of the tall-spiked habit of growth, begin to bloom below, following upward till they finish at the top.

Some sort of garden knowledge is so generally professed in these days, and so much more gardening of the better kind is being attempted, that people are gradually learning the advantage of planting in good groups of one thing at a time. The older way of putting one each of the same plant at regular intervals along a border—like buttons on a waistcoat—is now no longer tolerated, but a great deal has yet to be learnt. Even planting in bold groups, however good the plants, will be ineffective if not absolutely

unfortunate, if relationships of colouring are not understood. The safest plan is to plant in harmonies more or less graduated as to the warm colours, such as full yellow with orange and scarlet, and to plant blues with contrasts of yellows and any white flowers. Then delightful effects may be obtained with masses of grey foliage, such as Lavender, Lavender-cotton, and Stachys, and white Pink, with flowers that have colourings of tender pink, white, lilac and purple. To acquire a colour eye is an education in itself, founded on the needful natural aptitude, a gift that is denied to some people even if they are not actually colour-blind. But it is a precious possession where it occurs, and all the better

RAMSCLIFFE: LARKSPUR
FROM THE PICTURE IN THE POSSESSION OF MISS KENSIT

99

when it has been so well trained that the eye is enabled to appreciate the utmost refinements of colour-values, and when this education has been carried to the point necessary for the artist, of justly estimating the colour *as it appears to be*. This is the most difficult thing to learn; to see colour as it is, is quite easy; any one not colour-blind can do this; but to see it as it appears to be needs to be learnt, for upon this acquired proficiency depends the power of the artist to interpret the colours of objects and to represent them in their right relation to each other.

There is another good double flower-border in this pleasant garden. In the sunny month of August the fine Summer Daisies (*Chrysanthemum maximum*), Phloxes and Lavender are in beauty, and some bloom remains upon the climbing Roses. The Box-edging, stout and strong, can withstand the temporary encroachments of some of the border flowers, for in such a garden, rule is relaxed whenever such latitude tends to beauty. Here and there, where the little edging shrub showed signs of unusual vigour, it has been allowed to grow up on the understanding that it shall submit to the shears, which clip it into rounded ball-shapes of two sizes, one upon another, like loaves of bread.

A garden like this, of moderate size, and needing no troublesome accessories of glasshouses, or even frames, and very little outside labour, is probably the very happiest possession of its kind. As the seasons succeed each other new pictures of flower beauty are revealed in constant succession. After the day's work in the best of the daylight is over, its owner turns to it for pleasant labour or any such tending as it may need. Every group of plants meets him with a friendly face, for each one was

planted by his own hands. His watchful eye observes where anything is amiss and the needful aid is immediately given.

In a great garden this vigilant personal care of plants as individuals is impossible. However able a man the head gardener may be, or however much he may love and wish to cherish the flowers under his care, his duties and responsibilities are too many and too onerous to admit of his being able to enjoy this intimate fellowship; but in the humbler garden the close relationship of man and flowers, with all its beneficent and salutary serviceableness to both, seems to be exactly adjusted.

Such a garden it is that fulfils its highest purpose; that giving of the pleasure—the rich reward of the loving toil and care that have gone to its making; every plant or group in it doing its appointed work in its due season—that giving of "sweet solace" according to the well-fitting wording of our far-away ancestors.

And when the day's work is done, and the light just begins to fail, no one knows better than the artist that then is the best moment in the garden—when the colours acquire a wonderful richness of "subdued splendour" such as is unmatched throughout the lighter hours of the long summer day. Then it is that the flowers of delicate texture that have grown faint in the full heat, raise their heads and rejoice; that the tall evening Primrose opens its pale wide petals and gives off its faint perfume; that the little lilac cross-flowers of the night-scented Stock open out and show their modest prettiness and pour forth their enchanting fragrance. This early evening hour is indeed the best of all; the hour of loveliest sight, of sweetest scent, of best earthly rest and fullest refreshment of body and spirit.

LEVENS

FROM THE PICTURE IN THE POSSESSION OF Major Longfield

LEVENS

THERE is perhaps no garden in England that has been so often described or so much discussed as that at Levens in Westmorland, the home of Captain Jocelyn Bagot.

It was laid out near the beginning of the eighteenth century by a French gardener named Beaumont. There is nothing about it of the French manner, as we know it, for it is more in the Dutch style of the time, and has become in appearance completely English; according perfectly with the beautiful old house, and growing with it into a complete harmony of mellow age, whose sentiment is one of perfect unison both within and without.

Forward of the house-front, in a space divided by intersecting paths into six main compartments, is the garden. Flower-borders, box-edged on both sides, form bordering ornaments all round these divisions. The inner spaces are of turf. At the angles and at equal points along the borders are strange figures cut in yew and box. Some are like turned chessmen; some might be taken for adaptations of human figures, for one can trace a hat-covered head—one of them wears a crown—shoulders and arms and a spreading petticoat. Some of the yews, and these mostly in the more open spaces of grass or walk, rise four-square as solid blocks, with rounded roof and stemless mushroom finial. These have for the most part arched recesses, forming arbours. One of the tallest, standing clear on its little green, is differently shaped, being round in plan above and the stems bared all round below, with an encircling seat.

No doubt many of the yews have taken forms other than those that were originally designed; the variety of shape would be otherwise too daring; but these recklessly defiant escapes from rule only add to the charm of the place, presenting a fresh surprise at every turn. The play of light and variety of colour of the green surfaces of the clipped evergreens is a delight to the trained colour-eye. Sometimes in shadow, cold, almost blue, reflecting the sky, with a sunlit edge of surprising brilliancy of golden-green—often all bright gold-green when the young shoots are coming, or when the sunlight catches the surface in one of its many wonderful ways. For the trees, clipped in so many diversities of form, offer numberless planes and facets and angles to the light, whose play upon them is infinitely varied. Then the beholder, passing on and looking back, sees the whole thing coloured and lighted anew. This quantity of Yew and Box clipped into an endless variety of fantastic forms has often been criticised as childish. Would that all gardens were childish in so happy a way! Is not the joy and perfectly innocent delight that the true lover of flowers feels in a good garden in itself akin to childishness, and is not a fine old English garden such as this, with its numberless incidents that stir and gratify the imagination, and its abundance of sweet and beautiful flowers, just the one that can give that happiness in the greatest degree? Does not the oldest of our legends, so closely bound up with our youngest apprehension of religious teaching, tell us of the earliest of our race of whom we have any record or even tradition, living happily in a garden in a state of childish innocence? Why should a garden not be childish?—perhaps when it truly deserves such a term it is the highest praise it could possibly have!

However this may be the fact remains that those who own this garden of many wonders, and watch and tend it with unceasing love and reverence, and others who have had the happiness of working in it for many days together, find it a place that never wearies, but only continues day by day to disclose new beauties and new delights. Doubtless it is a garden that cannot be fairly judged from a hasty glance or a few hours' visit. Like many of the places and things that we call inanimate—though to one who knows and loves a garden nothing is more vitally living—such a place has its moods and can frown upon an unsympathetic beholder.

The garden is filled with many Roses and well-grown hardy plants;

LEVENS: ROSES AND PINKS
FROM THE PICTURE IN THE POSSESSION OF MRS. ARCHIBALD PARKER

those especially of tall stature making a fine effect. The Rose garden has White Pinks in its outer beds. Immediately beyond the garden's bounds is wild ground of a beautiful character. The river Kent, a rock-strewn stream with steep wooded banks, flows within fifty yards of the house. The contrast is a great and a delightful one. Wild parkland and untamed river without; and within the walls ordered restraint; then again, the quiet of the wide bowling-green, with its dark clipped hedges, and beyond it a long, tree-shaded walk.

Precious, indeed, are the few remaining gardens that have anything of the character of this wonderful one of Levens; gardens that above all others show somewhat of the actual feeling and temperament of our ancestors. They show personal discrimination combining happily with common-sense needs; walls and masses of yew and box to make shelter from the violence of wind, and yet to admit the welcome sunlight; so to provide the best conditions in the inner spaces for the growing of lovely flowers. Then the shaping of some of the yews into strange forms, shows perhaps the whimsical humour of some one of a line of owners, preserved, with careful painstaking, by his descendants.

A garden many generations old may thus be a reflection of the minds of several of such possessors—men who have not only thankfully paced its green spaces and delighted in its flowery joys, but who have held it in that close and friendly fellowship whose outcome is sure to be some living and lasting addition either to its comfort, its interest, or its beauty. The original design may have become in some degree lost, but unless the doings of the

several owners have been in the way of destruction or radical alteration, or something of obvious folly or bad taste, the garden will have gained in a remarkable degree that quality of human interest that is not easy to define but that is clearly perceptible, not only to a trained critic but to any one who has knowledge of its most vital needs and sympathy with its worthiest expression. This precious utterance is not confined to this or to any one special kind of gardening, but may pervade and illuminate almost any one of the many ways in which men find their pleasure and delight in ordering the sheltered seclusion of their home grounds, and enjoying the varied beauty of tree and bush and flower.

It is only in gardens of the most rigidly formal type, such as are full of architectural form and detail and admit of no alteration of the original plan, that personal influence can least be exercised. This is no doubt the reason why such gardens, correctly beautiful though they may be, are those that give in smallest measure that wonderful sense of the purest and most innocent happiness, that of all earthly enjoyments seems to be the most directly God-given.

Yet, even in such gardens, it is not impossible that some impress of the personal influence may be beneficently given, but the range of operation is extremely limited, the greatest knowledge and ability are needed, with the sure action of the keenest and most restrained judgment.

CAMPSEY ASHE

IN Eastern Suffolk, within a few miles of the sea, is this, the country home of the Hon. William Lowther.

The house, replacing an older one that occupied the same site, is of brick and stone, built in the earlier part of the nineteenth century. A moat, inclosing an unusually large area, and formerly entirely encompassing the house and garden, is now partly filled up; but one long arm remains, running the greater part of the length of the house and garden; a shorter length bounding the inclosed garden on the opposite side. The longer length of moat approaches the house closely on its eastern face, and then forms the boundary of a large and beautifully-kept square lawn, with fine old cedars and other trees. Following this southward is a double walled garden, with the main paths, especially those of the nearer division, bordered with flowers. Beyond these again is the portion of the garden that forms the subject of the picture—a small parterre of box-edged beds with a row of old clipped yews beyond. This leads westward to a grove of trees, with a statue also girt with trees standing in an oval in the midmost space.

The garden has beautiful incidents in abundance, but is somewhat bewildering. Traces of the older gardening constantly appear; but their original cohesion has been lost. The moat, always an important feature, ends suddenly at four points. Garden-houses and gazebos, that usually come at salient points with determinate effect, seem to have strayed into their places. Sections of the park seem to have broken loose and lost themselves in the garden. The garden is not the less charming in detail, but

is impossible to gather together or hold in a clear mental grasp, from the absence of general plan.

Besides the old clipped yews in the picture, others, apparently of the same age, inclose an oval bowling-green. In form they are as if they had been at first cut as a thick hedge with a roof-like sloping top. From this, at fairly regular intervals, spring great rounded masses, that, with the varying vigour of the individual trees and the continual clipping without reference to a fixed design, have asserted themselves after their own fashion. Though symmetry has been lost, the place has gained in pictorial value. Four ways lead in; the larger bosses guarding the entrances.

So it is throughout this charming but puzzling garden. Ever a glimpse of some delightful old-world incident, and then the baffled effort to fit together the disjointed members of what must once have been a definite design.

The portion of the garden that is simplest and clearest is a broad walk opposite the house, on the further side of the moat, and raised some ten feet above it; backed by an old yew hedge some twenty feet high, of irregular outline. Just opposite the middle of the house the line of the hedge is interrupted to give a view into the park, with a vista between groups of fine elms; but the hedge stretches away southward the whole length of the long arm of the moat and the walled gardens. At regular intervals along the old hedge are ranged, on column-shaped pedestals, busts that came from an Italian villa. About half way along steps lead down to the moat, where there is a ferry-punt propelled by an endless rope, such as is commonly used in the fenlands. At the end of the long

walk is a curious seat with a high carved back, that looks as if it had once formed part of an old ship or state barge, in the bygone days of two hundred years ago, when a fine style of bold and free wood-carving was lavishly used about their raised poops and stern-galleries.

Towards the end of the second division of the walled garden is an old orangery or large garden house, that probably was in connexion with the scheme of the yew hedges. It has the usual piercing with large lights but no top-light. The original purpose of these buildings was the housing of orange and other tender trees in tubs, and the fact of its presence might possibly throw some light on the mystery of the garden's former planning.

THE YEW HEDGE, CAMPSEY ASHE
FROM THE PICTURE IN THE POSSESSION OF
Mr. H. W. Search

Good hardy flowers are everywhere in abundance. Specially beautiful in the later summer is a grand pink Hollyhock of strong free habit, with the flowers of that best of all shapes—with wide, frilled outer petals and centres not too tightly packed.

It would be interesting work for some one with a knowledge of the garden design of the past three centuries in England to try to reconstruct the original plan of some one time. Though on the ground the various remaining portions of the older work cannot be pieced together, yet, if these were put on paper to proper scale, it might be possible to come to some general conclusions as to the way in which the garden was originally, and again perhaps subsequently, laid out. Some of the remaining portions of the older work of quite different dates may now seem to be of the same age, but the expert would probably be able to discriminate. The result of such a study would be worth having even if actual reconstruction were not contemplated.

CLEEVE PRIOR

NEAR a quiet village in Warwickshire, and in close relation to its accompanying farm buildings, is this charming old manor house. It is not upon a main road, but stands back in its own quiet place on rising ground above the Avon. Everything about it is interesting and quite unspoilt. The wooden hand-gate, with its acorn-topped posts, that stands upon two semi-circular steps may not have been of the pattern of the original gate— it has an eighteenth-century look—but it is just right now. It leads into a half dark, half light, double arcade of splendid old clipped yews. Looking from the gate they seem to be tall walls of yew to right and left, showing the projecting porch of the house at the end; but, passing along, there are seen to be openings between every two trees, each of which gives a charming picture of the lawns and simple flower beds to right and left. The path is paved with stone flags; the garden is bounded with a low wall of the local oolite limestone that rock-plants love. A few thin-topped old fruit-trees, their stems clothed with ivy, are another link between the past and present, and the somewhat pathetic evidence of their having long passed their prime and being on the downward path, is in striking contrast with the robust vigour of the ancient yews, already some centuries old, and looking as if they must endure for ever.

Eight yews stand on either side—sixteen in all. They are known as the twelve Apostles and the four Evangelists. The names may have belonged to them from the time of their planting, for the whole place belonged in old days to Evesham Abbey, and is pervaded with monastic memory and tradition. This may also account for the excellence of the

THE TWELVE APOSTLES, CLEEVE PRIOR
FROM THE PICTURE IN THE POSSESSION OF
SIR FREDERICK WIGAN

buildings, for the old monks were grand constructors, and their structures were not only solid but always beautiful.

One of the older of these at Cleeve Prior is a large circular dovecote of stone masonry with tiled roof and small tiled cupola. Such buildings were not unfrequent in the old days, and many of them remain. Sometimes they are round in plan, sometimes four-, sometimes eight-sided. Occasionally there is a central post inside, set on pivots to revolve easily, with lateral

arms carrying a ladder that reaches nearly to the walls, so that any one of the many pigeon-holes can be reached.

To the left of the Apostles' Garden, as you stand facing the house, a little gate leads into the vegetable garden. It has narrow grass paths bordered with old-fashioned flowers. A further gate leads into the orchard. Behind the house is the home close with some fine trees; on two other sides are the farm buildings, yard and rickyard.

How grandly the flowers grow in these old manor and farm gardens! How finely the great masses of bloom compose, and how beautifully they harmonise with the grey of the limestone wall and the wonderful colour of the old tiled roof; both of them weather and lichen-stained; each tile a picture in itself of grey and orange and tenderest pink.

The yews have got over their paler green colour of the early summer when the young shoots are put forth, and have settled into the deep green dress that they will wear till next May. For the time is September; wheat harvesting is going on and the autumn flowers are in full vigour. There are Dahlias, the great annual Sunflowers and the tall autumn Daisy; Lavender and Michaelmas Daisies, with sweet herbs for the kitchen, just as it should be in such a garden.

Some of these old pot-herbs are beautiful things deserving a place in any flower garden. Sage—for instance—a half shrubby plant with handsome grey leaf and whorled spikes of purple flowers; a good plant both for winter and summer, for the leaves are persistent and the plant well clothed throughout the year. Hyssop is another such handsome thing, of the same family, with a quantity of purple bloom in the autumn, when it is a great

favourite with the butterflies and bumble bees. This is one of the plants that was used as an edging plant in gardens in Tudor days, as we read in Parkinson's "Paradisus," where Lavender-cotton, Marjoram, Savoury and Thyme are also named as among the plants used for the same purpose. Rue, with its neat bluish-green foliage, is also a capital plant for the garden where this colour of leafage is desired. Fennel, with its finely-divided leaves and handsome yellow flower, is a good border plant, though rarely so used, and blooms in the late autumn. Lavender and Rosemary are both so familiar as flower-garden plants that we forget that they can also be used as neat edgings, if from the time they are young plants they are kept clipped. Borage has a handsome blue flower, as good as its relation the larger *Anchusa.* Tansy, best known in gardens by the handsome *Achillea Eupatorium,* was an old inmate of the herb garden. Sweet Cicely (*Myrrhis odorata*) has beautiful foliage, pale green and fern-like, with a good umbel of white bloom, and is a most desirable plant to group with and among early blooming flowers. And we all know what a good garden flower is the common Pot Marigold.

The old farm buildings at Cleeve Prior are scarcely less beautiful than the manor-house itself, and are remarkable for the timber erections, open at the sides but with tiled roofs, that give sheltered access, by outside stairways, to the lofts.

Throughout England the older farmhouses and buildings are full of interest, not only to architects, but to many who are in sympathy with good and simple construction, and have taken the pleasant trouble to learn enough about it to understand how and why the buildings were reared.

And in these restless days of hurry and strain and close competition in trades, and bad, cheap work, it is good to pass a quiet hour in wandering about among structures set up four or even five centuries ago by these grand building monks. The present writer had just such a pleasure not long ago in the South of England, where a large group of monastic farm buildings stands within sound of the wash of the sea. They are on sloping ground, inclosing three sides of a square; a wall, backed with trees, forming the fourth side. On the upper level is a great barn; a much greater, the tithe barn, being opposite it on the lower. Buildings containing stables, cattle-sheds and piggeries connect the two. Between these and the wall opposite is a spacious yard; across the middle is a raised causeway dividing the yard into two levels.

CLEEVE PRIOR: SUNFLOWERS
FROM THE PICTURE IN THE POSSESSION OF
Mr. James Crofts Powell

The barns are of grand masonry. Some of the stones, next above the plinth—a feature that adds so much to the dignity of the building, and by its additional width, to its solidity—measure as much as four feet six inches in length by twenty inches in height. In every fifth course is a row of triangular holes for ventilation, such as every brick or stone-built barn must have. They are cleverly arranged as to the detail of the manner of their building, and though only intended for use have a distinctly ornamental value. Where the walls rise at the gable ends they are corbelled out at the eaves and carried up some two feet above the line of the rafters, finishing in a wrought stone capping, thus stopping the thatch. For the buildings are, and always have been, thatched with straw, the ground around being good corn-land, a rich calcareous loam.

There is a delightful sense of restfulness about these fine solid buildings, built for the plainest needs of the community of the material nearest to hand, in the simply right and therefore most beautiful way. With no intentional ornament, they have the beauty of sound, strong structure and unconsciously right proportion. There is also a satisfaction in the plain evidence of delight in good craftsmanship, and in the unsparing use of both labour and material.

CONDOVER

CONDOVER HALL near Shrewsbury is a stately house of important size and aspect—one of the many great houses that were reared in the latter half of the reign of Queen Elizabeth. Its general character gives the impression of severity rather than suavity, though the straight groups of chimneys have handsome heads, and the severe character is mitigated on the southern front by an arcade in the middle space of the ground floor. The same stern treatment pervades the garden masonry. No mouldings soften the edges of the terrace steps; parapets and retaining walls, with the exception of the balustrade of the main terrace, are without ornament of light and shade; plainly weathered copings being their only finish. Only here and there, a pier that carries a large Italian flower-pot has a little more ornament of rather massive bracket form.

The garden spaces are large and largely treated, as befits the place and its environment of park-land amply furnished with grand masses of tree-growth. On the southern side of the house, where the ground falls away, are two green flats and slopes, leading to a lower walk parallel with their length and with the terrace above. The steps in the picture are the top flight of a succession leading to these lower levels. The lower and narrower grassy space has a row of clipped yews of a rounded cone-shape. The upper level has a design of the same, but of different patterns.

The balustrade in the picture is old, probably of the same date as the house; much of the other stonework is modern. The circular seat on a raised platform, with its stone-edged flower-beds, has a very happy effect, and its yew-hedge backing joins well into the older yews that

CONDOVER: THE TERRACE STEPS
FROM THE PICTURE IN THE POSSESSION OF
MISS AUSTEN LEIGH

overlap the parapet of the steps; their colour contrasting distinctly with that of the more distant Ilex, a magnificent example of a tree that deserves more general use in English gardens. The parterre above the steps and on a level with the house has box hedges, after the Italian manner, three feet high and two feet wide. These, with some of the yew hedges, were planted a hundred years ago, though much of the garden, with its ornaments of fine Italian flower-pots, was the work of the former owner, the late Mr. Reginald Cholmondeley, a man of powerful personality and fine taste.

The most important part of the garden lies to the west of the house, where there is a double garden of stiff pattern with high box borders and clipped evergreens. At a right angle to this, the spectator, standing at some distance westward, and looking back towards the east and straight with the space between the pair of gardens of angular design, sees a broad space flanked on either side by a row of handsome upright yews. The ground between is a flower garden of large diamond-shaped beds in two sizes, with cleverly-arranged green edgings. But now that the large Irish yews have grown to their early maturity, dominating the garden and insisting on their own strong parallel lines, it is open to question whether it would not have been better to have had a wide, clear middle space of green straight down the length, with the flowers in shapely, ordered masses to right and left. The close succession of large beds gives the impression of impediments to comfortable progress.

It was wise to leave the Irish yews unclipped. Though the common English yew is the tree that is of all others the most docile to the discipline of training and shearing, the upright growing variety will have none of it. In some fine English gardens they are clipped, always with disastrous effect. They will only take one form: that of an ugly swollen bottle, or lamp-chimney with a straight top. Their own form is quite symmetrical enough for use in any large design.

SPEKE HALL

THERE are, alas! but few now remaining of the timber buildings of the sixteenth century that are either so important in size or so well-preserved as this beautiful old Lancashire house.

They were built at a time when the country had settled down into a peaceable state; when houses need no longer be walled and loopholed against the probable raids of enemies; when their windows might be of ample size and might look abroad without fear. Many of them, however, were still moated, for a moat was of use not only for defence but as a convenient fish-pond, and as a bar to the depredations of wolves, foxes and rabbits.

The advance of civilisation also brought with it a greater desire for home comforts, and the genius of the country, unspoilt, unfettered, undiluted by that mass of half-digested knowledge of many styles that has led astray so many of the builders of modern days, by a natural instinct cast these dwellings into forms that we now seek out and study in the effort to regain our lost innocence, and that in many cases we are glad to adopt anew as models of what is most desirable for comfort and for the happy enjoyment of our homes.

Still, in these days we cannot build such houses anew without a suspicion of strain or affectation. When they were reared, oak was the building material most readily to be obtained, and carpenters' work, already well developed in the construction of roofs, now given free scope in outer walls as well, seemed to revel in the new liberty, and oak-framed houses

grew up into beautiful form and ornament in such a way as has never been surpassed in this country.

SPEKE HALL
FROM THE PICTURE IN THE POSSESSION OF
MR. GEORGE S. ELGOOD

It was satisfying and beautiful because every bit of ornamental detail grew out of the necessary structure. The plainer framing of cottage and farmhouse became enriched in the manor-house into a wealth of moulding and carving and other kinds of decoration. External panel ornament gained a rich quality by the repetition of symmetrical form, while the overhanging of the successive stories and the indentations between projecting wings and porches threw the various faces of the building into interesting masses of light and shade. Then, in delightful and restful

contrast to the "busy" wall-spaces, are the roofs, with their long quiet lines of ridge and their covering of tile or stone, painted by the ages with the loveliest tinting of moss and lichen.

Within, these fine old wooden houses show the good English oak as worthily treated as without. For the whole structure is of wood from end to end, built as soundly and strongly as were the old wooden ships. The inner walls, where they were not panelled with oak, were hung with tapestry. Ceilings of the best rooms were wrought with plaster ornament; lesser rooms showed the beams and often the thick joists that fitted into them and upheld the floor above. Where, as was usual, there was a long gallery in the topmost floor, its ceiling would show a tracery of oak with plaster filling, partly following the line of the roof. The whole structure, blossoming out in its more important parts into beautiful decorative enrichment, showed the worker's delight in his craft, and his mastery of mind and hand in conceiving and carrying out the possibilities offered by what was then the most usual building material of the country.

Such another house as Speke is Moreton Old Hall in Cheshire, but the latter is still more richly decorated, with carved strings, some of which were painted, and wood and plaster panels of great elaboration, and lead-quarried windows of large size and beautiful design.

The destruction of large numbers of these timber buildings in the eighteenth century can never be sufficiently deplored. There was a time when the fashion for buildings of classical form was spreading over England, when they were considered barbarous relics of a bygone age, and

when the delightful gardens that had grown up around them were alike condemned and in many cases destroyed.

There is not a large garden at Speke, but just enough of simple groups of flowers to grace the beautiful timber front. The picture shows that the gardening is just right for the place; not asserting itself overmuch but doing its own part with a restful, quiet charm that has a right relation to the lovely old dwelling.

GARDEN ROSES

THOSE who follow the developments of taste in modern gardening, cannot fail to perceive how great has been the recent increase in the numbers of Roses that are for true beauty in the garden.

It is only some of the elders among those who take a true and lively interest in their gardens who know what a scarcity of good things there was thirty years ago, or even twenty, compared with what we have now to choose from. Still, of the Roses commonly known as garden Roses, there were even then China Roses, Damask, Cabbage and Moss, Sweetbriars and Cinnamon Roses, and the free-growing Ayrshires, which are even now among the most indispensable.

But the wave of indifferent taste in gardening that had flooded all England with the desire for summer bedding plants, to the almost entire exclusion of the worthier occupants of gardens, had for a time pushed aside the older garden Roses. For whereas in the earlier half of the nineteenth century these good old Roses were much planted and worthily used, with the coming of the fashion for the tender bedding plants they fell into general disuse; and, with the accompanying neglect of many a good hardy border plant, left our gardens very much the poorer, and, except for special spring bedding, bare of flowers for all the earlier part of the year.

Now we have learnt the better ways, and have come to see that good gardening is based on something more stable and trustworthy than any passing freak of fashion. And though the foolish imp fashion will always pounce upon something to tease and worry over, and to set up on a temporary pedestal only to be pulled down again before long, so also it

assails and would make its own for a time, some one or other point of garden practice. Just now it is the pergola and the Japanese garden; and truly wonderful are the absurdities committed in the name of both.

But the sober, thoughtful gardener smiles within himself and lets the freaks of fashion pass by. If he has some level place where a straight covered way of summer greenery would lead pleasantly from one quite definite point to another, and if he feels quite sure that his garden-scheme and its environment will be the better for it, and if he can afford to build a sensible structure, with solid piers and heavy oak beams, he will do well to have a pergola. If he has travelled in Japan, and lived there for some time and acquired the language, and has deeply studied the mental attitude of the people with regard to their gardens, and imbibed the traditional lore so closely bound up with their horticultural practice, and is also a practical gardener in England—then let him make a Japanese garden, if he will *and can*; but he will be the wiser man if he lets it alone. Even with all the knowledge indicated, and, indeed, because of its acquirement, he probably would not attempt it. When a Japanese garden merely means a space of pleasure-ground where plants, natives of Japan, are grown in a manner suitable for an English garden, there is but little danger of going wrong, but such danger is considerable when an attempt is made to garden in the Japanese manner.

This is a wide digression from the subject of garden Roses, and yet excusable in that it can scarcely be too often urged that any attempt to practise anything in horticulture for no better reason than because it is the fashion, can only lead to debasement and can only achieve futility.

Now that there are large numbers of people who truly love their gardens, and who show evidence of it by giving them much care and thought and loving labour, the old garden Roses have been sought for and have been restored to their former place of high favour. And our best nurserymen have not been slow to see what would be acceptable in well-cared-for gardens throughout the length and breadth of the land; so that the last few years have seen an extraordinary activity in the production of good Roses for garden effect. The free-growing *Rosa polyantha* of the Himalayas has been employed as a seed or pollen-bearing parent, and from it have been developed first the well-known Crimson

"VISCOUNTESS FOLKESTONE"
FROM THE PICTURE IN THE POSSESSION OF
MR. R. CLARKE EDWARDS

Rambler, and later a number of less showy but much more refined flowers of just the right kind for free use in garden decoration.

Valuable hybrids have also been raised from the Tea Roses, one of the best known of them being *Viscountess Folkestone*, the subject of the picture; a grand Rose for grouping in beds or clumps, and one that yields its large, loose, blush-white flowers abundantly and for a long season. This merit of an extended blooming season runs through the greater number of the now long list of varieties of the beautiful hybrid Teas.

Some of the new seedling Tea Roses have nearly single flowers, and are none the less beautiful, as those wise folk well know who grow *Corallina* and the lovely white *Irish Beauty*, and its free-blooming companion *Irish Glory*. These also are plants that will succeed, as will most of the hybrid Teas, in some poor hot soils where most Roses fail.

Then for rambling over banks we have *Rosa wichuraiana* and its descendants; among these the charming *Dorothy Perkins*, good for any free use.

Those who garden on the strong, rich loams that Roses love will find that many of the so-called show Roses are grand things as garden Roses also; indeed, for purely horticultural purposes there is no need of any such distinction. The way is for a number of Roses to be grown on trial, and for a keen watch to be kept on their ways. It will soon be seen which are those that are happiest in any particular garden, and how, having regard to their colour and way of growth, they may best be used for beauty and delight.

In the garden where the picture was painted, *Viscountess Folkestone* has an undergrowth of Love-in-a-mist, that comes up year after year, and with its

quiet grey-blue colouring makes a charming companionship with the faint blush of the Roses.

PENSHURST

THE gardens that adorn the ancient home of the Sidneys are, as to the actual planting of what we see to-day, with repairs to the house and some necessary additions to fit it for modern needs, the work of the late Lord de L'Isle with the architect George Devey, begun about fifty years ago. It was a time when there was not much good work done in gardening, but both were men of fine taste and ability, and the reparation and alteration needed for the house, and the new planting and partly new designing of the garden could not have been in better hands.

The aspect and sentiment of the garden, now that it has grown into shape—its lines closely following, as far as it went, the old design—are in perfect accordance with the whole feeling of the place, so that there seems to be no break in continuity from the time of the original planting some centuries ago. Such as it is to-day, such one feels sure it was in the old days—in parts line for line and path for path, but throughout, just such a garden as to general form, aspect, and above all, sentiment, as it must have been in the days of old. For when it was first planted the conditions that would have to be considered were always the same; requirement of shelter from prevailing winds; questions relating to various portions, as to whether it would be desirable to welcome the sunlight for the flowers' delight, or to shut it out for human enjoyment of summer coolness—all such grounds of motive were, just as now, deliberated by the men of old days, whose decisions, actuated by sympathy with both house and ground, would bring forth a result whose character would be the same, whether thought out and planned to-day or four centuries ago.

So it is that we find the old work at Penshurst confirmed and renewed,

"GLOIRE DE DIJON," PENSHURST
FROM THE PICTURE IN THE POSSESSION OF
Sir Reginald Hanson Bart.

and new work added of a like kind, such as will make use of the wider modern range of garden plants, while it retains the dignity and grandeur of the fine old place.

The house stands on a wide space of grass terrace commanding the garden. On a lower level is a large quadrangular parterre, with cross paths. In each of its square angles is a sunk garden with a five-foot-wide verge of turf and a bordering stone kerb forming a step. The beds within, filled with good hardy plants, have bold box edgings eighteen inches high and a foot thick, that not only set off the bright masses of flowers within, but have in themselves an air of solidity and importance that befits the large scale of the place. They represent in their own position and on their lesser scale somewhat of the same character as the massive yew hedges, twelve feet high and six feet through, that do their own work in other parts of the garden.

These grand yew hedges and solid box borders have responded well to good planting and tending, for the late Lord de L'Isle knew his work and did it thoroughly. Not only was the ground well prepared, but for several years after planting the young trees were provided with a surface dressing that prevented evaporation and provided nutriment. This was carefully attended to, not only in the case of the yews but of the box edgings also.

The cross-walks of the parterre do not meet in the middle, but sweep round a circular fountain basin, in the centre of which stands a statue of what may be a young Hercules, brought from Italy by Lord de L'Isle. The slender grace of the figure might at first suggest a youthful Bacchus, but the identity in such a statue is easily established by looking for one or

other of the characteristic attributes of Hercules; these usually are the lion's skin, the upright-growing hair on the forehead, the poplar wreath or the battered, flattened ears. But the statue stands too far from the walk to be exactly identified.

That the nearer portions of the garden are on the same lines as the older planting is shown by an engraving in Harris's "Kent," where the parterre is, now as then, bounded by terraces on two of its sides, the house side and that of the adjoining churchyard, to which access is gained by a beautiful gabled gateway of brick and stone, the work of Tudor times.

The old churchyard has its own beauty, while the church and a fine group of elms are seen from the garden above the wall, and take their own beneficent place in the garden landscape.

The rectangular fountain, which, with its surrounding yew hedges, and the grass walks also inclosed by thick yew hedges, divides the two portions of the kitchen garden, are also parts of the old design, added to by the late owner. The yew hedges beyond the fountain pool have been set back to allow width enough for a handsome flower-border on either side. Water Lilies grow in the pool and the flower-borders display their beauties beyond, while the fruit trees of the kitchen garden show above the thick green hedges as flowering masses in spring, and in later summer, as the taller perennials of the border rise to their full height, as a thin copse of fruit and leafage. The turf walk and flower-border swing outward to suit the greater width of another fountain-basin at the end. This has straight sides running the way of the main path, and a segmental front. Instead of the usual rising kerb, there are two shallow stone steps, the upper one

even with the grass, the lower half way between that and the water-level. Except that it is less of a protection than something of the parapet kind, this is a most desirable means of near access to the water; welcome to the eye in all ways and allowing the water-surface to be seen from a distance. It is pleasantly noticeable in this pool that the water-level rises to the proper place. Nothing is more frequent or more unsightly than a deep pool or basin with straight sides and only a little water in the bottom. If the height of the water is necessarily fluctuating it is a good plan to build the tank in a succession of such steps; they are pleasant to see both above and under water, and in the case of an accident to a straying child, danger is reduced to the smallest point.

The picture shows one of the flights of steps from one level to another. To the left two handsome gate-piers and a fine wrought-iron gate lead to a quiet green meadow. Near by and just across it is the Medway, with wooded banks and groups of fine trees. The old wall is beautiful from the meadow side; its coping a garden of wild flowers. Above it is seen the clipped yew hedge with its series of rising ornaments, rounded in the direction of the axis of the hedge, but flat on

THE TERRACE STEPS, PENSHURST
FROM THE PICTURE IN THE POSSESSION OF
MR. FREDERICK GREENE

its two faces. This is seen in the picture on the upper level, above the steps to the left.

Herbaceous plants are grandly grown throughout this beautiful garden. Specially noticeable are the fine taste and knowledge of garden effect with which they have been used. There are not flowers everywhere, but between the flowery portions of the garden are quiet green spaces that rest and refresh the eye, and that give both eye and brain the best possible preparation for a further display and enjoyment of their beauty.

Such an example the picture shows. On either side is a border with masses of strong-growing hardy plants—pale Monkshood, Evening Primrose, Sweet-William, Pink Mallow—then, above the steps, only the restful turf underfoot, and to right and left the quiet walls of yew; at the end a group of great elms. At the foot of the steps, passing away to the right, is another double flower-border, passing again by a turn to the right into the quiet green walk leading to the large fountain basin.

Many a good climbing Rose, with other rambling and clambering plants, find their homes on the terraces. A *Gloire de Dijon* or one of its class—*Madame Bérard* or *Bouquet d'Or*, perhaps; either of these the equal of the other for such garden use—rises from below the parapet of one of the flights of steps and comes forward in happiest fellowship with a leaden vase of fine design; the dark background of Irish Yew making the best possible ground for both Rose and urn.

In olden days these lead ornaments were commonly painted and gilt, but the revived taste for all that is best in gardening rightly considers such treatment to be a desecration of a surface which with age acquires a beautiful grey colouring and a delightful quality of colour-texture. The painting of lead would seem to be a relic of the many toy-like artifices in gardening that were prevalent in Tudor times. All these are rejected in the best modern practice, though all the old ways that made for true garden beauty and permanent growth and value have been retained.

A clever way of utilising the stronger growing Clematises, including the large purple Jackmanii, is here practised. They are swung garland-fashion between a row of Apple-trees that borders one of the walks. Hops are

used in the same way. It was perhaps a remembrance of Italy, where Vines are trained to swing between the Mulberries.

The beautiful pale yellow Carnation, named *Pride of Penshurst*, was raised in this good garden, where everything tells of the truest sympathy with all that is best in English horticulture. Not the least among the soothing and satisfying influences of the pleasure-grounds of Penshurst, is the entire absence of the specimen conifer, that, with its wearisome repetition of single examples of young firs and pines, has brought such a displeasing element of restless confusion into so many pleasure-grounds.

BRICKWALL

EAST SUSSEX is rich in beautiful houses of Tudor times; many precious relics remaining of those days and of the Jacobean reigns, of important manor-house, fine farm building and labourer's cottage. These were the times when English oak, some of the best of which grew in the Sussex forests, was the main building material. The walls were framed of oak, and the same wood provided beams, joists and rafters, boards for the floors, panelling, doors, window-mullions and furniture. In those days the wood was not cut up with the steam-saw, but was split with the axe and wedge. The carpentry of the roofs was magnificent; there was no sparing of stuff or of labour. Much of it that has not been exposed to the weather, such as these roof-framings, is as sound to-day as when it was put together, with its honest tenons and mortises and fastenings of stout oak pins. In most cases the old oak has become extraordinarily hard and of a dark colour right through.

Brickwall, the beautiful home of the Frewens, near Northiam, is a delightful example, both as to house and garden, of these old places of the truest English type. A stately gateway, and a short road across a spacious green forecourt bounded by large trees, leads straight to the entrance in the wide north-eastern timbered front. The other side of the house, in closely intimate relation to the garden, has a homely charm of a most satisfying kind.

The wide bricked path next the house, so typical of Sussex, speaks of the strong, cool soil. The ground rises just beyond, and again further away in the distance. The garden is divided into two nearly equal portions by the

double flower-border, backed by pyramidal clipped yews, that forms the subject of the picture, and is enfiladed by the middle windows of the house. In the left hand division is a long rectangular pool with rather steep-angled sides of grass, looking a little dangerous. There is a reason here for the water being a good way below the level of the lawn, for this is much above that of the ground floor of the house; still it is a matter of doubt whether it would not have been better to have had a flagged or bricked path some four feet wide, not much above the water-level, with steps rising on two sides to the lawn, and dry walling, allowing of some delightful planting, from the path to the lawn level.

On the two outer sides of the garden, parallel with the middle walk, are raised terraces, reached by steps at the ends of the bricked path. These have walls on their outer sides, and towards the garden, yew hedges kept low so that it is easy to see over. These low hedges run into a much higher and older one that connects them towards the upper end of the garden.

The clipped yews which give the garden its character are for the most part of one pattern, a tall three-sided pyramid, only varied by some tall cones. One cannot help observing how desirable it is in gardens of this kind that the form of the clipped yews should for the most part keep to one shape, or at any rate one general pattern, just as the architecture, whatever its character or ornament, within some kind of limit remains faithful to the dominant idea.

The picture shows a double flower-border in August dress; good groups of the best hardy plants combining happily with some of the pyramids of yew. To the right is the fine summer Daisy (*Chrysanthemum maximum*), with

142

the lilac Erigeron and the spiky blue-purple balls of the Globe Thistle, the tall double Rudbeckia Golden Glow, Lavender, Poppies and Phlox. To the left, Phloxes and the tall Evening Primrose, the great garden Tansy (*Achillea Eupatorium*), seed-heads of the Delphiniums that bloomed a month ago, White Mallow, and the grand red-ringed Sweet-William called Holborn Glory. Everything speaks of good cultivation on a rich loamy soil, for those fine yews want plenty of nutriment themselves, and would also be apt to rob their less robust neighbours. But then the good gardener knows how to provide for this.

There is always an opportunity for beautiful treatment, when, as in

BRICKWALL, NORTHIAM
FROM THE PICTURE IN THE POSSESSION OF
MR. R. A. OSWALD

143

this case, the garden ground ascends from the house. The garden is laid out to view, almost as a picture hangs on a wall, in the very best position for the convenience of the spectator; and there is nothing that gives a greater sense of dignity, with something of a poetical mystery, than separate flights of steps ascending one after another in plane after plane— as they do in that magnificent example, Canterbury Cathedral. It matters not whether the steps are under a roof or not—the impression received is the same. And there is much beauty in the steps themselves being long and wide and shallow. Looking uphill we see the steps; looking downhill they are lost. It is not the foot only that rests upon the step, it is the eye also, and that is why any handsome steps with finely-moulded edges are so pleasant to see. The overhanging edge may have arisen from utility, in that, where a step must be narrow it gives more space for the foot; but in the wide step it affords still more satisfaction, giving a good shadow under the moulded edge, and accentuating the long level lines that are so welcome to the eye.

STONE HALL, EASTON

THE FRIENDSHIP GARDEN

IT was a pleasant thought, that of the lover of good flowers and firm friend of many good people, who first had the idea of combining the two sentiments into a garden of enduring beauty.

Such a garden has been made at Easton by the Countess of Warwick. The site of the Friendship Garden has been happily chosen, close to the remains of an ancient house called Stone Hall, which now serves Lady Warwick as a garden-house and library of garden books.

The flower-plots are arranged in a series of concentric circles; the plants are the gifts of friends. The name of each plant and that of the giver are recorded on an imperishable majolica plaque. Many well-known givers' names are here, from that of the very highest in the land downward. The plants themselves comprise many of the best and handsomest.

The picture shows the garden as it is about the middle of September; the time of the great White Pyrethrum, the perennial Sunflowers and the earliest of the Michaelmas Daisies. The bush of Lavender is blooming late, its normal flowering time is a month earlier. But Lavender, especially when some of the first bloom is cut, will often go on flowering, as later-formed shoots come to blooming strength. Let us hope that the giver is not shortlived like the gift, for Lavender bushes, after a few years of strong life, soon wear out. Already this one is showing signs of age, and it would be well to set a few cuttings in spring or autumn, or, still better, to layer it by one of the lower branches, in order to renew

STONE HALL, EASTON: THE FRIENDSHIP GARDEN
FROM THE PICTURE IN THE POSSESSION OF
THE COUNTESS OF WARWICK

the life of the plant when the strength of the present bush comes to an end.

Such a garden, full of so keen a personal interest, sets one thinking. What will become of it fifty or a hundred years hence? The flowers, with due diligence of division, and replanting and enriching of the soil, will live for ever. The name-plates, with care and protection from breakage, will also live. But what will these names be one or two generations hence? Will the plants all be there? And what of the Friendships? They are something

147

belonging intimately to the lives of those now living. What record of them will endure; or enduring, be of use or comfort to those who come after?

Then one thinks and wonders—what hand, perhaps quite a humble one—planted the old apple-tree that has its stem now girdled by a rustic seat. Its days are perhaps already numbered; the top is thin and open, the foliage is spare; it seems to be beyond fruit-bearing age, and as if it had scarcely strength to draw up the circulating sap.

And then, for all the carefulness given to the making of the garden and its tender memories of human kindness in giving and receiving, the plant that dominates the whole, and gives evidence of the oldest occupation from times past, and promises the greatest attainment of age in days to come, is the Ivy on the old Stone Hall. Probably it was never planted at all—came by itself, as we carelessly say—or planted, as we may more thoughtfully and worthily say, by the hand of God, and now doing its part of sheltering and fostering the Garden of Friendship. Should not the Ivy also have its heart-shaped plate and its most grateful and reverent inscription, as a noble plant, the gift of the kindest Friend of all, who first created a garden for the sustenance and delight of man and put into his heart that love of beautiful flowers that has always endured as one of the chiefest and quite the purest of his human pleasures?

There is a rose-garden beyond the bank of shrubs to the left, where each Rose, on one of the permanent labels, here shaped after the pattern of a Tudor Rose, has a quotation from the poets. Here are, among others, the older roses of our gardens, the Damask and the Rose of Provence, the

Cinnamon and the Musk Rose, the bushy Briers and the taller Eglantine that we now call Sweetbrier.

Close at hand there is also a Shakespeare Garden, designed to show what were the garden-flowers commonly in use in his time. Here we may again find Rosemary—that sweetly aromatic shrub, so old a favourite in English gardens. Its long-enduring scent made it the emblem of constancy and friendship. And here should be Rue, also classed by Shakespeare among "nose-herbs," and the sweet-leaved Eglantine, and Lads-Love, Balm and Gilliflowers (our Carnations), a few kinds of Lilies, Musk and Damask Roses, Violets, Peonies, and many others of our oldest garden favourites.

THE DEANERY GARDEN, ROCHESTER

THOSE who know the Dean of Rochester,[A] either personally or by reputation, will know that where he dwells there will be a beautiful garden. His fame as a rosarian has gone throughout the length and breadth of Britain, and far beyond, and his practical activity in spreading and fostering a love of Roses must have been the means of gladdening many a heart, and may be reckoned as by no means the least among the many beneficent influences of his long and distinguished ministry.

[A] These lines were in print before the lamented death of Dean Hole.

A few days' visit to Dean Hole's own home at Caunton Manor, near Newark, will ever remain among the writer's pleasantest memories. It must have been five and twenty years ago, and it was June, the time of Roses. To one whose home was on a poor sandy soil it was almost a new sight to see the best of Roses, splendidly grown and revelling in a good loam. Not that the credit was mainly due to the nature of the garden ground, for, as the Dean (then Canon Hole) points out in his delightful "Book about Roses," the soil had to be made to suit his favourite flower. In this, or some one of his books, he feelingly describes how many of the visitors to his garden, seeing the splendid vigour of his Roses, at once ascribed it to the excellence of his soil. "Of course," they said, "your flowers are magnificent, but then, you see, you have got such a soil for Roses." "I should think I had got a soil for Roses," was the reply, "didn't I mix it all myself and take it there in a barrow?" I quote from memory, but this is the sense of this excellent lesson. The writer's own experience is exactly the same. Of the quantities of garden visitors who have come—their number

has had to be stringently limited of late—not one in twenty will believe that one loves a garden well enough to take a great deal of trouble about it. In fact, it is only this unceasing labour and care and watchfulness; the due preparation according to knowledge and local experience; the looking out for signal of distress or for the time for extra nourishment, water, shelter or support, that produces the garden that satisfies any one with somewhat of the better garden knowledge; a knowledge that does not make for showy parterres or for any necessarily costly complications; rather, indeed, for all that is simplest, but that produces something that is apparent at once to the eye, and sympathetic to the mind, of the true garden-lover.

It must have been a painful parting from the well-loved Roses and the many other beauties of the Caunton garden, when the new duties of honourable advancement called Canon Hole from the old home to the Deanery of Rochester; from the pure air of Nottinghamshire to that of a town, with the added reek of neighbouring lime and cement works. But even here good gardening has overcome all difficulties, and though, when the air was more than usually loaded with the foul gases given off by these industries, the Dean would remark, with a flash of his characteristic humour, that Rochester was "a beautiful place—to get away from," yet the Deanery garden is now full of Roses and quantities of other good garden flowers, all grandly grown and in the best of health. Roses are in fact rampant. A rough trellis, simply made of split oak after the manner of the hurdles used for folding sheep in the Midlands, but about six feet high, stands at the back of the main double flower-border. Rambling Roses and others of free-growing habit are loosely trained to this, their great heads of

bloom hanging out every way with fine effect; each Rose is given freedom to show its own way of beauty, while the trellis gives enough support and guides the general line of the great hedge of Roses.

The Dean is not alone among the flowers, for Mrs. Hole is also one of the best of gardeners.

The picture shows a portion of a double flower-border where a curving path connects two others that are at different angles. In the

THE DEANERY GARDEN, ROCHESTER
FROM THE PICTURE IN THE POSSESSION OF
MR. G. A. TONGE

distance, rising to a height of a hundred feet, is the grand old Norman keep; the rare Deptford Pink (*Dianthus Armeria*) grows in its masonry. The ancient city wall is one of the garden's boundaries. Another old wall, that is within the garden, has been made the home of many a good rock-plant. On the left, in the picture, are masses of Poppies, Roses and White Lilies, with Alströmeria, Love-in-a-Mist, and Larkspurs, both annual and perennial; the background is of the soft, feathery foliage of Asparagus. The Roses are of all shapes; single and double; show Roses and garden Roses; standards, bushes and free-growing ramblers. On the right are more Larkspurs, Irises in seed-pod, Lavender, and some splendidly-grown Lilium szovitsianum, one of the grandest of Lilies, and, where it can be grown like this, one of the finest things that can be seen in a garden. Its tender lemon colouring has suffered in the reproduction, which makes it somewhat too heavy.

The upper part of a greenhouse shows in the picture. It is sometimes impossible to keep such a structure out of sight, but one like this, of the plainest possible kind, is the least unsightly of its class. It is just an honest thing, for the needs of the garden and for a part of its owner's pleasure. The fatal thing is when an attempt is made to render greenhouses ornamental, by the addition of fretted cast-iron ridges and fidgety finials. These ill-placed futilities only serve to draw attention to something which, by its nature, cannot possibly be made an ornament in a garden, while it is comparatively harmless if let alone, and especially if the wood-work is not painted white but a neutral grey. In all these matters of garden structures; seats, arbours and so forth, it is much best in a simple garden to keep to

what is of modest and quiet utility. In the case of a large place, which presents distinct architectural features, it is another matter; for there such details as these come within the province of the architect.

COMPTON WYNYATES

IN the very foremost rank among the large houses still remaining that were built in Tudor times is the Warwickshire home of the Marquess of Northampton. The walls are of brick, wide-jointed after the old custom, with quoins, doorways, and window-frames of freestone, wrought into rich and beautiful detail in the heads of the bays and the grand old doorway, whose upper ornament is a large panel bearing the sculptured arms of King Henry VIII.

Formerly the house was entirely surrounded by a moat, which approached it closely on all sides but one, where a small garden was inclosed. Now, on the three sides next the building, grass lawns take its place. On all sides but one, hilly ground rises almost immediately; in steep slopes for the most part, beautifully wooded with grand elms.

To the north is the small garden still inclosed by the moat. Straight along it is a broad grass walk with flower borders on both sides, leading to a thatched summer-house that looks out upon the moat. Lesser paths lead across and around among vegetables and old fruit-trees. At one corner is a venerable Mulberry.

The space within the quadrangle of the building is turfed and has cross-paths paved with stone flags. Bushes of hardy Fuchsia mark their outer angles of intersection. At the foot of the walls hardy Ferns are in luxuriance, and nothing could better suit the place. There are a few climbing Roses, but they are not overdone; the beautiful building is sufficiently graced, but not smothered, by vegetation. So it is throughout the place both within and without; house and garden show a loving

COMPTON WYNYATES
FROM THE PICTURE IN THE POSSESSION OF
Mr. George S. Elgood

reverence for the grand old heritage and that sound taste and knowledge that create and maintain well and wisely.

From the portions of the site of the old moat that are now grass, a turf slope rises to a height of about eight feet. On the upper level is a gravel walk, and beyond it a yew hedge about four feet high, with ornaments of peacocks cut in it at the principal openings, and of ball and such-like forms at other apertures. This is on the level of the main parterre. A wide gravel path divides the garden into two equal portions, swinging round in the middle space to give place to a circular grass-plot with a sundial.

This beautiful place offers so few details that can be adversely criticised that these few are the more noticeable. The sundial has a handsome shaft, but should stand upon a much wider step. The introduction of pyramid fruit-trees at concentric points, both here and in other parts of the design, is an experiment of doubtful value, that will probably never add to the pictorial value of the design. The garden critic may also venture to suggest that the pergola, which is well placed at the eastern end of the parterre, deserves better piers than its posts of fir. Here would be the place for some simple use of specially made bricks, such as a pier hexagonal in plan built of bricks of two shapes, diamond and triangle, two inches thick, with a wide mortar joint. Each course would take two bricks of each shape, and their disposition, alternating with each succeeding course, would secure an admirable bond.

The great parterre has main divisions of grass paths twelve feet wide, each subdivision—four on each side of the cross-walk and sundial—of eight three-sided beds disposed Union-Jack-wise, with bordering beds stopped by a clipped Box-bush at each end. Narrow grass walks are between the beds. The borders are roomy enough to accommodate some of the largest of the good hardy flowers, for the garden is given to these, not to "bedding stuff." Here are some of the tall perennial Sunflowers, eight feet high; the great autumn Daisy (*Pyrethrum uliginosum*); bushes of Lavender with Pentstemons growing through them—a capital combination, doing away with the need of staking the Pentstemons; the last of the Phloxes; for the time of the picture, which was painted in this part of the garden, is September.

This bold use of autumnal border flowers invites the exercise of invention and ingenuity; for instance, August is the main time for the flowering of Lavender, and, though a thinner crop succeeds the heavier normal yield, yet the bushes then look thin of bloom. Clematises, purple, lilac and white, can be planted among them, and can easily be guided, by an occasional touch with the hand, to run over the Lavender bushes. The same capital autumn flowers should be planted with the handsome white Everlasting Pea. The Pea is supported by stout branching spray and does its own good work in July. When the bloom is over it is cut off, and the Clematis, which has been growing by its side with a support of rather slighter spray, is drawn close to the foliage of the Pea and spreads over it.

The working out of such simple problems is one of the many joys of the good gardener; and every year, with its increased experience, brings with it a greater readiness in the invention of such happy combinations.

CHINA ROSES AND LAVENDER, PALMERSTOWN
FROM THE PICTURE IN THE POSSESSION OF
MRS. KENNEDY-ERSKINE

PALMERSTOWN

THE Earl of Mayo's residence in County Kildare, Ireland, lies a few miles distant from the small market town of Naas. The house is of classical design, built within the nineteenth century. Around it is extensive parkland that is pleasantly undulating, and is well furnished with handsome trees both grouped and standing singly. In the lower level is a large pool with Water-lilies, and natural banks fringed with reeds and the other handsome sub-aquatic vegetation that occurs wild in such places.

There was an older house at Palmerstown in former days, whose large walled kitchen garden remains. It is a lengthy parallelogram, divided in the middle into two portions, each nearly a square, by a fine old yew hedge with arches cut in it for the two walks that pass through. The paths are broad, and some width on each side has been planted as a flower border, giving ample space for the good cultivation and enjoyment of all the best of the hardy flowers, so willing to show their full growth and beauty in the soft genial climate of the sister island.

It is a place that shows at once the happy effect of wise and sure direction, for Lady Mayo is an accomplished gardener, and the inclosure abounds with evidences of fine taste and thoughtful intention. One length of border is given to Lavender and China Rose, always a delightful and most harmonious mixture. There is a length of some twenty yards of this pleasant combination—the picture shows one end—with a few groups of taller plants, such as Bocconia, behind. Fruit-trees, trained as espaliers, form the back of the border, or sometimes there is a hedge of Sweet Pea. Vegetables occupy the middle portions of the quarters. The flower-

bordered paths pass across and across the middle space, with others about ten feet within the walls and parallel with them. Quite in the middle the path passes round a fountain basin, and there are four arches on which Roses and Clematis are trained.

Such flower borders give ample opportunity for the practising of good gardening. The task is the easier in that only one of the pairs of borders can be seen at a glance, and a definite scheme of colour progression can easily be arranged. Such schemes are well worth thinking out. The writer's own experience favours a plan in which the borders begin with tender colourings of pale blue, white and pale yellow, with bluish foliage, passing on to the stronger yellows. These lead to orange, scarlet and strong blood-reds. The scale of colouring then returns gradually to the pale and cool colours.

It is by such simple means that the richest effects of colour are obtained, whether in a continuous border or in clump-shaped masses. A separate space of flower-border may also be well treated by the use of an even more restricted scheme of colouring. Purple and lilac flowers, with others of pink and white only, and foliage of grey and silvery quality, the darkest being such as that of Rosemary and Echinops, make a charming flower-picture, with a degree of pictorial value that any one who had not seen it worked out would scarcely think possible.

The right choice of treatment depends in great measure on the environment. When this, as at Palmerstown, consists of old walls and a grand hedge of venerable yews, a suitable frame is ready for the display of almost any kind of garden-picture.

The yews are ten feet high and six feet through. Over a seat one of them is cut into the form of a peacock. To the left of the green archway in the Lavender picture, the yew takes the form of the heraldic wild-cat, the Mayo crest. Outside the garden is a yew walk of untrimmed trees; they show in the picture to the right, over the wall. Here, in the heat of summer, the coolness and dim light are not only in themselves restful and delightful, but, after passing along the bright borders, where eye and brain become satiated with the brilliancy of light and colour, the cool retreat is doubly welcome, preparing them afresh for further appreciation of the flower-borders.

ST. ANNE'S, CLONTARF

THERE is perhaps no place within the British islands so strongly reminiscent of Italy as St. Anne's, in County Dublin, one of the Irish seats of Lord Ardilaun. This impression is first received from the number and fine growth of the grand Ilexes, which abound by the sides of the approaches and in the park-land near the house. For there are Ilexes in groups, in groves, in avenues—all revelling in the mild Irish air and nearness to the sea.

The general impression of the place, as of something in Italy, is further deepened by the house of classical design and of palatial aspect, both within and without, that has that sympathetic sumptuousness that is so charming a character of the best design and ornamentation of the Italian Renaissance. For in general when in England we are palatial, we are somewhat cold, and even forbidding. We stand aloof and endure our greatness, and behave as well as we are able under the slightly embarrassing restrictions. But in Italy, as at St. Anne's, things may be largely beautiful and even grandiose, and yet all smiling and easily gracious and humanly comforting.

As it is in the house, so also is it in the garden; the same sentiment prevails, although the garden shows no effort in its details to assume an Italian character. But apparent everywhere is the remarkable genius of Lady Ardilaun—a queen among gardeners. A thorough knowledge of plants and the finest of taste; a firm grasp and a broad view, that remind one of the great style of the artists of the School of Venice—these are the acquirements and cultivated aptitudes that make a consummate gardener.

The grounds themselves were not originally of any special beauty. All the present success of the place is due to good treatment. Adjoining the house, at the northern end of its eastern face, is a winter garden. Looking from the end of this eastward, is a sight that carries the mind directly to Southern Europe; an avenue of fine Ilexes, and, at the end, a blue sea that might well be the Mediterranean. Passing to the left, before the Ilexes begin, is a walk leading into a walled inclosure of about two acres. Next the wall all round is a flower-border; then there is a space of grass, then a middle group of four square figures, each bordered on three sides by a grand yew hedge that is clipped into an outline of enriched scallops like the edge of a silver *Mentieth*; a series of forms consisting of a raised half-circle, then a horizontal shoulder, and then a hollow equal to the raised member.

The genial climate admits of the use of many plants that generally need either winter housing or some special contrivance to ensure well-being. Thus there are great clumps of the blue African Lily (*Agapanthus*); and Iris Susiana blooms by the hundred, treated apparently as an ordinary border plant.

The picture shows a portion of a double flower-border in another square walled garden, formerly a kitchen garden, and only comparatively lately converted into pleasure-ground. Yews planted in a wide half-circle form a back-ground to the bright flowers and to a statue on a pedestal. The intended effect is not yet finished, for the trellis at the back of the borders is hardly covered with its rambling Roses, which will complete the picture by adding the needed height that will bring it into proper relation with the

tall yews. There is a cleverly invented edging which gives added dignity as well as regularity, and obviates the usual falling over of some of the contents of the border on to the path; an incident that is quite in character in a garden of smaller proportion, but would here be out of place. A narrow box edging, just a trim line of green, has within it a good width of the foliage of Cerastium. The bloom, of course, was over by the middle of June, but the close carpet of downy white leaf remains as a grey-white edging throughout the summer and autumn.

Though this border shows bold masses of flowers, it scarcely gives an

ST. ANNE'S, CLONTARF
FROM THE PICTURE IN THE POSSESSION OF
MISS MANNERING

idea of the general scale of grand effect that follows the carrying out of the design and intention of its accomplished mistress. For here things are done largely and yet without obtrusive ostentation. They seem just right in scale. For instance, in the house are some great columns; huge monoliths of green Galway marble. It is only when details are examined that it is perceived how splendid they are, and only when the master tells the story, that the difficulty of transporting them from the West of Ireland can be appreciated. For they were quarried in one piece, and bridges broke under their immense weight. At one point in their journey they sank into a bog, and their rescue, and indeed their whole journey and final setting up at its end, entailed a series of engineering feats of no small difficulty.

AUCHINCRUIVE

THE mild climate of south-western Scotland is most advantageous for gardening. Hydrangeas and Myrtles flourish, Fuchsias grow into bushes eight to ten feet high. Mr. Oswald's garden lies upon the river Ayr, a few miles distant from the town of Ayr. The house stands boldly on a crag just above the river, which makes its music below, tumbling over rocky shelves and rippling over shingle-bedded shallows. For nearly a mile the garden follows the river bank, in free fashion as befits the place. Trees are in plenty and of fine growth, both on the garden side and the opposite river shore. Here and there an opening in the trees on the further shore shows the distant country. The garden occupies a large space; the grouping of the shrubs and trees taking a wilder character in the portions furthest away from the house, so that, mingling at last with native growth, garden gradually dies away into wild. Large undulating lawns give a sense of space and freedom and easy access.

That close, fine turf of the gardens of Britain is a thing so familiar to the eye that we scarcely think what a wonderful thing it really is. When we consider our flower and kitchen gardens, and remember how much labour of renewal they need—renewal not only of the plants themselves, but of the soil, in the way of manurial and other dressings; and when we consider all the digging and delving, raking and hoeing that must be done as ground preparation, constantly repeated; and then when we think again of an ancient lawn of turf, perhaps three hundred years old, that, except for moving and rolling, has, for all those long years, taken care of itself; it

seems, indeed, that the little closely interwoven plants of grass are things of wonderful endurance and longevity. The mowing

AUCHINCRUIVE

FROM THE PICTURE IN THE POSSESSION OF MR. R. A. OSWALD

prevents their blooming, so that they form but few fresh plants from seed. Imperceptibly the dying of the older plants is going on, and the hungry root-fibres of their younger neighbours are feeding on the decaying particles washed into the earth.

But whether lawns could exist at all without the beneficent work of the earthworms is very doubtful. Every one has seen the little heaps of worm-castings upon grass, but it remained for Darwin, after his own long

experiment and exhaustive observation, partly based upon and comprehending the conclusions of other naturalists, to tell us how largely the fertility of our surface soil is due to the unceasing work of these small creatures. Worms swallow large quantities of earth and decaying leaves, and Darwin's observations led him "to conclude that all the vegetable mould over the whole country has passed many times through, and will again pass many times through the intestinal canals of worms." This, indeed, is the only way in which it is possible for a person with any knowledge of the needs of plant-life, to conceive the possibility of any one closely-packed crop occupying the same space of ground for hundreds of years. The soil, as it passes, little by little, through the bodies of the worms, undergoes certain chemical changes which fit it afresh for its ever-renewed work of plant-sustenance.

There are some who, viewing the castings as an eyesore on their lawns, cast about for means of destroying the worms. This is unwise policy, and would soon lead to the impoverishment of the grass. The castings, when dry, are easily broken down by the roller or the birch-broom, and the grass receives the beneficent top-dressing that assures its well-being and healthy continuance.

The only part of the garden at Auchincruive that is obedient to rectangular form, is the kitchen garden and the ground about it. The kitchen garden lies some way back from the house and river, and, with its greenhouses, is for the most part hidden by two long old yew hedges which run in the direction of the river. One of these appears in the picture, with its outer ornament of bright border of autumn flowers. Here are Tritomas,

Gypsophila in mist-like clouds, tall Evening Primrose and Campanula pyramidalis, both purple and white, with many other good hardy flowers.

The red-leaved tree illustrates a question which often arises in the writer's mind as to whether trees and shrubs of this coloured foliage, such as Prunus Pissardi and Copper Beech and Copper Hazel are not of doubtful value in the general garden landscape. Trees of the darkest green, as this very picture shows by its dark upright yews, are always of value, but the red-leaved tree, though in the present case it has been tenderly treated by the artist, is apt to catch the eye as a violent and discordant patch among green foliage. Especially is this the case with the darker form of copper beech, which, in autumn, takes a dull, solid, heavy kind of colour, especially when seen from a little distance, that is often a disfiguring blot in an otherwise beautiful landscape.

The same criticism may occasionally apply to trees of conspicuous golden foliage, but errors in planting these, though often made, may easily be avoided by suitable grouping and association with white and yellow flowers. Indeed it would be delightful to work out a whole golden garden.

THE YEW ARBOUR, LYDE
FROM THE PICTURE IN THE POSSESSION OF Mr. George E. B.
Wrey

YEW ARBOUR: LYDE

IT is not in large gardens only that hardy flowers are to be seen in perfection. Often the humblest wayside cottage may show such a picture of plant-beauty as will put to shame the best that can be seen at the neighbouring squire's. And where labouring folk have a liking for clipped yews, their natural good taste and ingenuity often turns them into better forms than are seen among the examples of more pretentious topiary work.

The cottager has the undoubted advantage that, as his tree is usually an isolated one, he can see by its natural way of growth the kind of figure it suggests for his clipping; whereas the gardener in the large place usually has to follow a fixed design. So it is that one may see in a cottage garden such a handsome example as the yew in the picture.

The lower part of the tree is nearly square in plan, with a niche cut out for a narrow seat. There is space enough between the top of this and the underside of the great mushroom-shaped canopy, to allow the upper surface of the square base to be green and healthy. The great rounded top proudly carries its handsome crest, that is already a good ornament and will improve year by year. The garden is raised above the road and only separated from it by a wall which is low on the garden side and deeper to the road. It passes by the side of the yew, so that the occupier of the seat commands a view of the road and all that goes along it, and can exchange greetings and gossip with those who pass by.

The cottagers of the neighbourhood—it is in Herefordshire, about four miles from Hereford—have a special fancy for these clipped yews; many

examples may be met with in an afternoon's walk. Not very far from this is a capital peacock excellently rendered in conventional fashion. It stands well above a high pedestal, one side of which is hollowed out for a little seat.

One may well understand what a pride and pleasure and amusing interest these clipped trees are to the cottage folk; how after each year's clipping they would discuss and criticise the result and note the progress of the growth towards the hoped-for form. A pile of cheeses is a favourite pattern, sometimes on a square base, with the topmost ornament cut into a spire or even a crown.

The English peasant has a love for ornament that always strives to find some kind of expression. In many parts the thatcher makes a kind of basket ornament on the top of his rick; and the pattern of crossed laths, pegged down with the hazel "spars" that finishes the thatched cottage roof near the eaves, is of true artistic value. The carter loves to dress his horses for town or market, and a fine team, with worsted ribbons in mane and headstall, and quantities of gleaming highly-polished brasses, is indeed a pleasant sight upon the country high road.

Now, alas! when cheap rubbish, misnamed ornamental, floods village shops and finds its way into the cottages, the cottager's taste, which was always true and good as long as it depended on its own prompting and instinct, and could only deal with the simplest materials, is rapidly becoming bewildered and debased. All the more, therefore, let us value and cherish these ornaments of the older traditions; the bright little gardens and the much-prized clipped yew.

A usual feature of these cottage yews is that the seat is for one person alone. The labourer sits in his little retreat enjoying his evening pipe after his day's work, while the wife puts the children to bed and gets the supper. Probably he has been harvesting all day, and his strong frame is tired, with that feeling of almost pleasant fatigue that comes to a wholesome body after a good day's work well done; and when the hardly-earned rest is thoroughly enjoyed. So he sits quite quiet, with one eye on the possible interests of his outer world, the road, and the other on the beauty of his flower-border. And what a pretty double border it is, with its grand mass of pink Japan Anemone and well-flowered clump of Goldilocks (one of the few yellow-bloomed Michaelmas Daisies), looking at its near relation in purple over the way.

A graceful little Plum-tree shoots up through the flowers; its long free shoots of tender green seem to laugh at the rigid surface of the clipped yew beyond. Don't be too confident about your freedom, little Plum; it is more than likely that you will be severely pruned next winter. But you need not mind, for if you lose one kind of beauty, you will gain others; the pure white bloom of spring-time, and the autumn burden of purple fruit; and be both handsome and useful, like your neighbour the old yew.

AUTUMN FLOWERS

HOW stout and strong and full of well-being they are—the autumn flowers of our English gardens! Hollyhocks, Tritomas, Sunflowers, Phloxes, among many others, and lastly, Michaelmas Daisies. The flowers of the early year are lowly things, though none the less lovable; Primroses, Double Daisies, Anemones, small Irises, and all the beautiful host of small Squill and Snow-Glory and little early Daffodils. Then come the taller Daffodils and Wallflowers, Tulips, and the old garden Peonies and the lovely Tree Peonies. Then the true early summer flowers.

If you notice, as the seasons progress, the average of the flowering plants advances in stature. By June this average has risen again, with the Sea Hollies and Flag Irises, the Chinese Peonies and the earlier Roses. And now there are some quite tall things. Mulleins seven and eight feet high, some of them from last year's seed, but the greater number from the seed-shedding of the year before; the great white-leaved Mullein (*Verboscum olympicum*), taking four years to come to flowering strength. But what a flower it is, when it is at last thrown up! What a glorious candelabrum of branching bloom! Perhaps there is no other hardy plant whose bulk of bloom on a single stem fills so large a space. And what a grand effect it has when it is rightly planted; when its great sulphur spire shows, half or wholly shaded, against the dusk of a wood edge or in some sheltered bay, where garden is insensibly melting into woodland. This is the place for these grand plants (for their flowers flag in hot sunshine), in company with white Foxglove and the tall yellow Evening Primrose, another tender bloom that is shy of sunlight. Four o'clock of a June morning is the time

175

to see these fine things at their best, when the birds are waking up, and but for them the world is still, and the Cluster-Roses are opening their buds. No one can know the whole beauty of a Cluster-Rose who has not seen it when the summer day is quite young; when the buds of such a rose as the Garland have just burst open and the sun has not yet bleached their wonderful tints of shell-pink and tenderest shell-yellow into their only a little less beautiful colouring of full midday.

By July there are still more of our tall garden flowers; the stately Delphiniums, seven, eight, and nine feet high; tall white Lilies; the tall yellow Meadow-Rues, Hollyhocks, and Sweet Peas in plenty.

By August we are in autumn; and it is the month of the tall Phloxes. There are some who dislike the sweet, faint and yet strong scent of these flowers; to me it is one of the delights of the flower year.

No garden flower has been more improved of late years; a whole new range of excellent and brilliant colouring has been developed. I can remember when the only Phloxes were a white and a poor Lilac; the individual flowers were small and starry and set rather widely apart. They were straggly-looking things, though always with the welcome sweet scent. Nowadays we all know the beauty of these fine flowers; the large size of the massive heads and of the individual blooms; the pure whites, the good Lilacs and Pinks, and that most desirable range of salmon-rose colourings, of which one of the first that made a lively stir in the world of horticulture was the one called *Lothair*. In its own colouring of tender salmon-rose it is still one of the best. Careful seed-saving among the brighter flowers of this colouring led to the tints tending towards scarlet, among which *Etna* was a

distinct advance, to be followed, a year or two later, by the all-conquering *Coquelicot*. Some florists have also pushed this docile flower into a range of colouring which is highly distasteful to the trained colour-eye of the educated amateur; a series of rank purples and virulent magentas; but these can be avoided. What is now most wanted, and seems to be coming, is a range of tender, rather light Pinks, that shall have no trace of the rank quality that seems so unwilling to leave the Phloxes of this colouring.

Garden Phloxes were originally hybrids of two or three North American species; for garden purposes they are divided into two groups, the earlier, blooming in July, much shorter in stature and more bushy, being known as the *suffruticosa* group, the later, taller kinds being classed as the *decussata*. They are a little shy of direct sunlight, though they can bear it in strong soils where the roots are always cool. They like plenty of food and moisture; in poor, dry, sandy soils they fail absolutely, and even if watered and carefully watched, look miserable objects.

But where Phloxes do well, and this is in most good garden ground, they are the glory of the August flower-border.

In the teaching and practice of good gardening the fact can never be too persistently urged nor too trustfully accepted, that the best effects are accomplished by the simplest means. The garden artist or artist gardener is for ever searching for these simple pictures; generally the happy combination of some two kinds of flowers that bloom at the same time, and that make either kindly harmonies or becoming contrasts.

In trying to work out beautiful garden effects, besides those purposely arranged, it sometimes happens that some little accident—such as the dropping of a seed, that has grown and bloomed where it was not sown—may suggest some delightful combination unexpected and unthought of. At another time some small spot of colour may be observed that will give the idea of the use of this colour in some larger treatment.

It is just this self-education that is needed for the higher and more thoughtful gardening, whose outcome is the simply conceived and beautiful pictures, whether they are pictures painted with the brush on paper or canvas, or with living plants in the open ground. In both cases it needs alike the training of the eye to observe, of the brain to note, and of the hand to work out the interpretation.

The garden artist—by which is to be understood the true lover of good flowers, who has taken the trouble to learn their ways and wants and moods, and to know it all so surely that he can plant with the assured belief that the plants he sets will do as he intends, just as the painter can

PHLOX AND DAISY
FROM THE PICTURE IN THE POSSESSION OF LADY MOUNT-
STEPHEN

compel and command the colours on his palette—plants with an unerring hand and awaits the sure result.

When one says "the simplest means," it does not always mean the easiest. Many people begin their gardening by thinking that the making and maintaining of a handsome and well-filled flower-border is quite an easy matter. In fact, it is one of the most difficult problems in the whole range of horticultural practice—wild gardening perhaps excepted. To achieve anything beyond the ordinary commonplace mixture, that is without plan or forethought, and that glares with the usual faults of bad colour-combinations and yawning, empty gaps, needs years of observation and a considerable knowledge of plants and their ways as individuals.

For border plants to be at their best must receive special consideration as to their many and different wants. We have to remember that they are gathered together in our gardens from all the temperate regions of the world, and from every kind of soil and situation. From the sub-arctic regions of Siberia to the very edges of the Sahara; from the cool and ever-moist flanks of the Alps to the sun-dried coasts of the Mediterranean; from the Cape, from the great mountain ranges of India; from the cool and temperate Northern States of America—the home of the species from which our garden Phloxes are derived; from the sultry slopes of Chili and Peru, where the Alströmerias thrust their roots deep down into the earth searching for the precious moisture.

So it is that as our garden flowers come to us from many climes and many soils, we have to bear in mind the nature of their places of origin the better to be prepared to give them suitable treatment. We have to know, for

instance, which are the few plants that will endure drought and a poor, hot soil; for the greater number abhor it; and yet such places occur in some gardens and have to be provided with what is suitable. Then we have to know which are those that will only come to their best in a rich loam, and that the Phloxes are among these, and the Roses; and which are the plants and shrubs that must have lime, or at least must have it if they are to do their very best. Such are the Clematises and many of the lovely little alpines; while to some other plants, many of the alpines that grow on the granite, and nearly all the Rhododendrons, lime is absolute poison; for, entering the system and being drawn up into the circulation, it clogs and bursts their tiny veins; the leaves turn yellow, the plant dies, or only survives in a miserably crippled state.

An experienced gardener, if he were blindfolded, and his eyes uncovered in an unknown garden whose growths left no soil visible, could tell its nature by merely seeing the plants and observing their relative well-being, just as, passing by rail or road through an unfamiliar district, he would know by the identity and growth of the wild plants and trees what was the nature of the soil beneath them.

The picture, then, showing autumn Phloxes grandly grown, tells of good gardening and of a strong, rich loamy soil. This is also proved by the height of the Daisies (*Chrysanthemum maximum*). But the lesson the picture so pleasantly teaches is above all to know the merit of one simple thing well done. Two charming little stone figures of *amorini* stand up on their plinths among the flowers; the boy figure holds a bird's nest, his girl companion a shell. They are of a pattern not unfrequent in English

gardens, and delightfully in sympathy with our truest home flowers. The quiet background of evergreen hedge admirably suits both figures and flowers.

It is all quite simple—just exactly right. Daisies—always the children's flowers, and, with them, another of wide-eyed innocence, of dainty scent, of tender colouring. Quite simple and just right; but then—it is in the artist's own garden.

MYNTHURST

AT the time the picture was painted, Mynthurst was in the occupation of Mrs. Wilson, to the work of whose niece, Miss Radcliffe, the garden owes much of its charm.

It lies in the pleasant district between Reigate and Dorking, on a southward sloping hill-side. The house is a modern one of Tudor character, standing on a terrace that has a retaining wall and steps to a lower level. The garden lies open to the south and south-westerly gales, the prevalent winds of the district, but it is partly sheltered by the walls of the kitchen garden, and by a yew hedge which runs parallel with one of the walls; the space so inclosed making a sheltered place for the rose garden. Here Roses rise in ranks one above the other, and have a delightful and most suitable carpet of Love-in-a-mist. This pretty annual, so welcome in almost any region of the garden, is especially pretty with Roses of tender colouring; whites, pale yellows, and pale pinks. A picture elsewhere shows it combined with Rose *Viscountess Folkestone.*

Beyond the rose garden, a path leads away at a right angle between the orchard and the kitchen-garden wall. Here is the subject of the picture. A broad border runs against the wall, as long as the length of the kitchen garden. A border so wide is difficult to manage unless it has a small blind alley at the back rather near the wall, to give access to what is on the wall and to the taller plants in the back of the border. But here it is arranged in another way. The front edge of the border is not continuous, but has little paths at intervals cutting across it and reaching nearly to the wall. This method of obtaining easy access also has its merits, though it involves a

large amount of edging. Mynthurst has a strong soil, an advantage not always to be had in this district, so that Roses can be well grown, and some of the Lilies. Here the Tiger Lily, that fine autumn flower, does finely. It is one of the Lilies that is puzzling, or as we call it, capricious, which only means that we gardeners are ignorant and do not understand its vagaries. For in some other heavy soils it refuses to grow, and in some light ones it luxuriates; but it is so good a plant that it should be tried in every garden.

It is a pretty plan to have the orchard in connexion with the flower-borders; though from the point of view of good gardening the wisdom is doubtful of having clumps of flowers round the trunks of the fruit-trees. Shallow-rooted annuals for a season or two may do no harm, but the disturbance of the ground needful for constant cultivation, with the inevitable consequence of worry and irritation of the fruit-trees' roots, can hardly fail to be harmful, though the effect meanwhile is certainly pretty. The evil may not show at once, but is likely to follow.

One does not often see so strong a Canterbury Bell in the autumn as the one in the picture. It must have been a weak or belated plant of last year that made strong growth in early summer. Sometimes one sees such a plant that had remained in the kitchen-garden reserve bed; left there because it was weaker than the ones taken for planting out in autumn. It is not generally known that these capital plants will bear potting when they are almost in bloom, so that when a few are so left, they can be used as highly decorative room plants, and have the advantage of lasting much longer than when in the open border, exposed to the sun. One defect

these good plants have, which is the way the dying flowers suddenly turn brown. Instead of merely fading and falling, and so decently veiling their decadence, the brown flowers hang on and are very unsightly. It is only, however, a challenge to the vigilance of the careful gardener; they must be visited in the morning garden-round and the dead flowers removed. It is like the care needed to arrest the depredations of the mullein caterpillar. It is no use wondering whether it will come, or hoping it will not appear; *it always comes* where there are mulleins, about the second week of June. When the first tiny enemy is seen, any mulleins there may be should be visited twice a day.

MYNTHURST
FROM THE PICTURE IN THE POSSESSION OF
MISS RADCLIFFE

In the front of the picture, just under the red rose, is a patch of *Mimulus*, one of the larger variations of the brilliant little *M. cardinalis*. All the kinds like a cool, strong soil; they are really bog plants, and revel in moisture. The old Sweet Musk, so favourite a plant in cottage windows, likes a half-shady place at the foot of a cool wall. Many a dull, sunless yard might be brightened by this sweet and pretty plant. The Welsh Poppy, with its bright pale-green leaves and good yellow bloom, is also excellent for the same use, but is best sown in place from a just-ripened pod.

ABBEY LEIX

IN a picturesque, but little-known district in Queen's County, Ireland, lies Abbey Leix, the residence of Lord de Vesci. It is a land of vigorous tree-growth and general richness of vegetation. Hedge-rows show an abundance of well-grown ash timber, and the park is full of fine oaks, a thing that is rare in Ireland, and that makes it more like English parkland of the best character. This impression is accentuated in spring-time when the oaks are carpeted with the blue of wild Hyacinths, and when the broad woodland rides are also rivers of the same Blue-bells.

In this favoured land the common Laurel is a beautiful tree, thirty feet high; the mildness of the winter climate allowing it to grow unchecked. Only those who have seen it in tree form in the best climates of our islands, or in Southern Europe, know the true nature of the Laurel's growth, or the poetry and mystery of its moods and aspects. The long grey limbs shoot upward and bend and arch in a manner almost fantastic. Sometimes a stem will incline downwards and run along the ground, followed by another. In the evening half-light they might be giant silver-scaled serpents, writhing and twisting and then springing aloft and becoming lost to sight in the dim masses of the crowning foliage. Seen thus one can hardly reconcile its identity with that of the poor, tamed, often-clipped bush of every garden. The Laurel is so docile, so easily coerced to the making of a quickly-grown hedge or useful screen, that its better qualities as an unmutilated tree in a mild district are usually lost sight of.

The house at Abbey Leix is a stone building of classical design of the middle of the eighteenth century. On the northern front is the entrance

ABBEY-LEIX
FROM THE PICTURE IN THE POSSESSION OF
SIR JAMES WHITEHEAD, BART.

forecourt; on the southern, the garden. Here, next the house, is a wide terrace, bounded on the outer side by the parapet of a retaining wall, and next the building, by a running *guilloche* of box-edged beds filled with low-growing plants. The terrace has a semi-circular ending, near the eastern wall of the house, formed of an evergreen hedge, with a wooden seat following the same line, and a sundial at the radial point. At the other end,

188

the terrace ends in a flight of downward steps leading to large green spaces, with fine trees and flowering shrubs, and eventually to the walled gardens. Straight across the terrace from the house is the parterre, whose centre ornament is an unusually well-proportioned fountain of the same date as the house. It is circular in plan, with a wide lower basin and two graduated superimposed tazzas. From this, four cross-paths radiate; the quarters are filled mainly with half-hardy flowers such as Gladiolus; the design being accentuated at several points by the upright growing Florence Court Yews. The parterre is inclosed by a low wall, backed by a clipped evergreen hedge; on the wall stand at intervals graceful stone figures of *amorini*, identical in character with those shown in the picture of Phlox and Daisy, and apparently designed by the same hand.

The steps at the western end of the terrace are wide and handsome, and are also ornamented with sculptured *amorini*. The path leads onward, at first directly forward, but a little later in a curved line through a region of lawn and stream, with trees and groups of flowering shrubs. Here and there, on the grass by itself, is one of the free-growing Roses, rightly left without any support, and showing the natural fountain-like growth that so well displays the beauty of many of the Roses of the old Ayrshire class and of some of the more modern ramblers. The path passes one end of an avenue of large trees, and, after a while, turning to the left, reaches the kitchen gardens, consisting of several walled inclosures. One of these, of which one wall is occupied by vineries, has been made into a flower garden, where hardy flowers, grandly grown, are in the wide borders next

189

the wall. A portion of such borders, in an adjoining compartment of the garden, forms the subject of the picture.

The inner space is divided into two squares, one having as a centre a rustic summer-house almost hidden by climbing plants; from this radiating grass paths pass between beds of flowers. The outer borders in the next walled compartment are ten feet wide, and are finely filled with all the best summer plants, perennial, annual and biennial. The fine pale yellow *Anthemis tinctoria* is here grown in the way this good plant deserves, and its many companions, Hollyhocks, Delphiniums, Japan Anemones, Phloxes and Lavender; annual Chrysanthemums, Gladiolus, Carnations, Tritomas, and all such good things, are cleverly and worthily used, and, with the graceful arches of free Roses and white Everlasting Pea, make delightful garden pictures in all directions.

The garden of Abbey Leix is one of those places that so pleasantly shows the well-directed intention of one who is in close sympathy with garden beauty; for everywhere it reflects the fine horticultural taste and knowledge of Lady de Vesci, who made the garden what it is.

MICHAELMAS DAISIES

EARLY in September, when the autumn flowers are at their finest, some of the Starworts are in bloom. Even in August they have already begun, with the beautiful low-growing *Aster acris*, one of the brightest of flowers of lilac or pale purple colouring. From the time this pretty plant is in bloom to near the end of October, and even later, there is a constant succession of these welcome Michaelmas Daisies. The number of kinds good for garden use is now so great that the growers' plant lists are only bewildering, and those who do not know their Daisies should see them in some good nursery or private garden and make their own notes. As in the case of Phloxes, the improvement in the garden kinds is of recent years, for I can remember the time when it was a rare thing to see in a garden any other Michaelmas Daisy than a very poor form of *Novi-Belgii*, a plant of such mean quality that, if it came up as a seedling in our gardens to-day, it would be sent at once to the rubbish heap.

When the learner begins to acquire a Daisy-eye he will see what a large proportion of the garden kinds are related to this same *Novi-Belgii*, the Starwort of New England. The greater number of the garden varieties are derived from North American species, but they hybridise so freely that it is now impossible to group the garden plants with any degree of botanical accuracy. But the amateur may well be content with a generally useful garden classification, and he will probably learn to know his *Novi-Belgii* first. Then he will come to those *Novi-Belgii* that are from the species *lævis*, rather wider and brighter green of leaf and only half the height. Then, once known, he cannot mistake *Novæ-Angliæ*, with its hairy and slightly

viscid stem and foliage, and strong smell, and its two distinct colourings—rich purples and reddish pinks. Then again, if he observes his plants in early summer, he can never mistake the heart-shaped root-leaves of *cordifolius* for any other. This is one of the most beautiful of the mid-season Starworts, with its myriads of small flowers gracefully disposed on the large spreading panicles. Of this the best known and most useful are *A. cordifolius elegans* and a paler-coloured and most dainty variety called *Diana*. Once seen he can never forget the low-growing early *A. acris* or the good garden varieties of *A. Amellus*, both from European species. Several other kinds, both tall and short, early and late, will be added to those named, but these may be taken as perhaps the best to begin with.

Where space can be given, it is well to set apart a separate border for these fine plants alone. This is done in the garden where Mr. Elgood found his subject. Here the Starworts occupy a double border about eight feet wide and eighty feet long. They are carefully but not conspicuously staked with stiff, branching spray cut out the winter before from oaks and chestnuts that had been felled. The spray is put in towards the end of June, when the Asters are making strong growth. The borders are planted and regulated with the two-fold aim of both form and colour beauty. In some places rather tall kinds come forward; in the case of some of the most graceful, such as *cordifolius Diana*, the growths being rather separated to show the pretty form of the individual branch. In others it was thought that their best use was as a flowery mass. Each kind is treated at the time of staking according to its own character, and so as best to display its natural form and most obvious use. Like all the best flower gardening it is the painting

of a picture with living plants, but, unlike painting, it is done when the palette is empty of its colours. Still the good garden-planter who has intimate acquaintance and keen sympathy with his plants, can plant by knowledge and faith; by knowledge in his certainty of recollection of the habit and stature and colour of his plants; by faith in that he knows that if he does his part well the growing thing will be docile to his sure guidance. In these borders of Michaelmas Daisies one other flowering plant is

MICHAELMAS DAISIES, MUNSTEAD WOOD
FROM THE PICTURE IN THE POSSESSION OF
MR. T. NORTON LONGMAN

admitted, and well deserves its place, namely, that fine white Daisy *Pyrethrum uliginosum*, otherwise *Chrysanthemum serotinum*. There can be no doubt that it is a daisy flower and that it blooms at Michaelmas; facts that alone would give it a right to a place among the Michaelmas Daisies. But it has all the more claim to its place among them in that it is the handsomest of the large white Daisies, and, though there are white kinds and varieties of the perennial Asters, not one of them can approach it for size or pictorial effect. There is also the still taller *Chrysanthemum leucanthemum* or *Leucanthemum lacustre*, but this is a plant that has an element of coarseness, and unless the spaces are large, and the Asters are thrown up to an unusual size by a strong and rich soil, it looks heavy and out of proportion.

Towards the front of the main portions of the Aster borders are rather bold, but quite informal edgings of grey-leaved plants such as white Pink, Stachys and Lavender-cotton; in places only a few inches wide, as where the rich purple, gold-eyed *Aster Amellus* comes to within a few inches of the path, in the white Pink's region, or again, where the grey, bushy masses of Lavender-cotton run in a yard deep among the Daisies.

About fifteen sorts are used in this double border; very early and very late ones are excluded, so as to have a good display from the third week of September for a month onward. They are mostly in rather large groups of one kind together.

There is a more than usual pleasure in such a Daisy garden, kept apart and by itself; because the time of its best beauty is just the time when the rest of the garden is looking tired and overworn—evidently dying for the year. Some trees are already becoming bare of leaves; the tall sunflowers look

bedraggled; Dahlias have been pinched by frost and battered by autumn gales, and it is impossible to keep up any pretence of well-being in the borders of other hardy flowers.

Then with the eye full of the warm colouring of dying vegetation and the few remaining blooms of perennial Helianthus and half-hardy marigolds of the fading borders, to pass through some screening evergreens to the fresh, clean, lively colouring of the lilac, purple and white Daisies, is like a sudden change from decrepit age to the brightness of youth, from the gloom of late autumn to the joy of full springtide.

Another excellent way of growing the perennial Asters is among shrubs, and preferably among Rhododendrons, whose rich green forms a fine background for their tender grace, and whose stiff branches give them the support they need.

THE ALCOVE, ARLEY
FROM THE PICTURE IN THE POSSESSION OF
MRS. CAMPBELL

ARLEY

THROUGHOUT the length and breadth of England it would be hard to find borders of hardy flowers handsomer or in any way better done than those at Arley in Cheshire. The house, an old one, was much enlarged by the late Mr. R. E. Egerton-Warburton, and the making of the gardens, now come to their young maturity, was the happy work of many years of his life. Here we see the spirit of the old Italian gardening, in no way slavishly imitated, but wholesomely assimilated and sanely interpreted to fit the needs of the best kind of English garden of the formal type, as to its general plan and structure. It is easy to see in the picture how happily mated are formality and freedom; the former in the garden's comfortable walls of living greenery with their own appropriate ornaments, and the latter in the grandly grown borders of hardy flowers.

The subject of the picture is the main feature in the garden plan. A path some fifteen feet wide, with grassy verges of ample width, and deep borders of hardy flowers. What is shown is about one fourth of the whole length. At the back of the right-hand border is the high old wall of the kitchen garden; on the left, as grand a wall of yew, ten feet high and five feet thick, its straight line pleasantly broken and varied by shaped buttresses of clipped yew, whose forms take that distinct light and shade, and strong variations of solidity of green colouring, that make the surfaces of our clipped English yew so valuable a ground-work for masses of brilliant flowers.

The same yew buttresses are against the wall on the right, placed symmetrically with the ones opposite. Near the end, as shown in the

picture, the last pair of buttresses come forward the whole width of the border, each buttress ending in an important shaped finial to the front. Between these and the well-designed alcove in stone masonry that so satisfactorily ends the walk, is a space of turf, leading on the left, through an arch cut in the ten-foot-high yew hedge, to the bowling-green. Nothing can make a more effective shelter than such grand yew hedges; the solid wall itself is scarcely better. Even on the roughest days, with a storm of wind of destructive power outside, the space within is calm and sheltered, and the flowers escape that cruel battering from fierce blasts that add so much to the difficulty of gardening in exposed places. But the planting and thus providing this much-needed shelter is just good gardening, and when, in addition, it is done to a design of happy invention and true proportion, with just such refinements of detail and ornament as are suited to the garden's calibre and the owner's endowment, then, with the addition of splendid masses of good flowers grandly grown, do we find gardening at its best.

The time of year of this picture is in or near the second week of July, when the White Lily is at its finest, and the Orange Lily is in bloom, with the Blue Delphinium and many another good garden flower. One can see how all the best garden flowers are utilised here. There is the White Sidalcea at the front of the border, one of the many plants of the Mallow family that are so important in our borders; for our grand Hollyhocks are Mallows too. This White Sidalcea has much the same value as the large White Snapdragon, one good variety of which, the precursor of the many good large kinds now grown, was the only one of its kind at the time the picture

was painted. Of late their numbers have greatly increased, and also their stature and the variety of their beautiful colourings, so that now they can be used as tall plants of great effect. Six feet two inches was the measurement of one grand spike of soft, rosy colouring in the writer's own garden last autumn. These capital plants have been "fixed," as gardeners say, in ranges of different heights; tall, intermediate, and the quite dwarf little cushions whose form is perhaps as little suited to the character of the plant as the foolish little dwarf Sweet Peas, that are only wilfully wanton, freakish distortions of a beautiful and graceful plant, whose duty it is to climb and bring its pretty blooms up to the level of our admiring eyes and appreciative noses. A good strong

THE ROSE GARDEN, ARLEY
from the picture in the possession of
MRS. HUTH

199

soil is shown by the well-being of the White Lily and Phlox, Sweet Williams and double Scarlet Potentilla. Carnations are largely grown in the borders; the great Orange Lily (*L. croceum*) has just given place to the White; Canterbury Bells are in grand masses, and the sturdier plants are interspersed with graceful fragilities, such as the long-spurred yellow Californian Columbine.

To the left of the alcove an archway cut in the yew hedge leads to the bowling-green. This also is inclosed and sheltered by yew hedges. There is a terrace all round, from which it is pleasant to watch the game. Next to this, and following along the line of the yew hedge, is a square inclosure of turf, with a few clipped yews. This is a kind of ante-room to the rose-garden. High walls of yew are all around except to this garden, where they are low and shaped. The middle space of the rose-garden has beds concentrically arranged, leaving spandrils of beds of other shape. At the end is a garden-house, and a wide way out to lawn spaces with fine trees and flowering shrubs. A broad gravel walk at the boundary of the lawn, with a wide grass outer verge and the knee-high top of the wall of a sunk fence, that separates it from the park, leads leftwards to the house. From this walk there is a very beautiful view across the steeply-falling gradient of the park to the lake. The park has grand old oak trees that fall into picturesque groups. Beyond the lake again are fine masses of timber. The lake is a sheet of water that takes a winding course and disappears among the trees.

The kitchen-garden walls are interesting survivals of an old way of treating fruit-trees. They are three feet thick and honeycombed with flues for

heating. It was a clumsy and unmanageable expedient practised in the days before the circulation of water in pipes heated from one boiler was understood. The modern orchard-house is much more convenient and its working absolutely under control.

The kitchen garden lies between the house and the newer gardens that have been described. The maze should not be forgotten. It is at the back of the alcove and the bowling-green. These old garden toys are very seldom planted now. Perhaps people have not time for them. Also they are costly of labour; the area of green wall of a maze of even moderate size, that has to be clipped yearly, if computed would amount to an astonishing figure. Now that the possibilities of other forms of garden delight are so much widened, it is small wonder that the maze should have fallen into disuse. It must have been amusing in the older days when people's lives were simpler and more leisured; but there are puzzles and difficulties enough in our more complicated days, and the influences that we now want in a garden are soothing tranquillities rather than bewildering perplexities. Near the maze and alcove is a group of three great Lombardy Poplars that tells with extremely fine effect from many parts of the garden. On one side of the house is an old parterre of the kind now but seldom seen out of Italy; with elaborate scrolls and arabesques of clipped box; the more characteristically Italian form of the "knotted" gardens of our Tudor ancestors. The English patterns were much nearer akin to those used so lavishly on gala clothing in the form of needlework of cording and braiding, and the strap-work of wood-carving, while the Italian parterre designs were drawn more freely in flowing lines and less rigid forms.

Opposite the porch is a sundial, supported by a kneeling figure of a black slave, of the same design as the one in the gardens of the Inner Temple, that was formerly at Clement's Inn, and is known as the "Blackamoor." Like this one the figure is of lead.

LADY COVENTRY'S NEEDLEWORK
from the picture in the possession of
MRS. APPLETON

LADY COVENTRY'S NEEDLEWORK

THIS is a pretty Midland name for the good garden plant commonly called Red Valerian, or Spur Valerian (*Centranthus ruber*), that groups so well in the picture with the straw-thatched beehives. How the name originated cannot be exactly stated, but may easily be inferred. There are several estates in the Midland Counties belonging to the Coventry family, and, bearing in mind what we know of the home life of our great-great-grandmothers of the late eighteenth century, it may be assumed that some Lady Coventry of that date was specially fond of the pretty needlecraft so widely practised among the ladies of that time.

Delightful things they are, these old needlework pictures, with a character quite different from that of their predecessors of Jacobean times. These were much stiffer in treatment and usually had figures; a lady and gentleman and a dog being usual subjects, and trees looking like those out of a Noah's Ark, no doubt interpretations of the stiffly-cut yew and box trees of the gardens of the same times.

But the workers of the flower-pictures of a hundred years later, and into the first quarter of the nineteenth century, for the most part chose flowers alone for their subjects. Sometimes a drawing was made, but many of them look as if they were worked direct from the flowers. It would appear that the worker would begin in the spring, with a Hyacinth; then would come Anemones, Tulips, Auriculas, Lilac, Roses and Lilies; a jumble of seasons but a concord of pretty things, and all done with a simplicity, a sweetness, a directness of intention and absence of strain or affectation, that give them a singular charm. One such picture that I have before me

must have been begun in May, and finished, perhaps, in August and September; for the first flower in the upper left-hand corner, where the work would naturally begin, is a thyrse of Lilac, and the last, low down on the right, is a Nasturtium; while the intermediate flowers, following each other in what would be approximately their natural sequence, come in between. These are Pansy, Rose, Sweet Pea, Love-in-a-Mist, Lily, Larkspur, Convolvulus, Carnation, Jasmine and Passion-flower; and one Daisy-shaped flower, whose identity, considering the numbers of possible Composites and the somewhat vague manner of the rendering, cannot be determined, though all the other flowers are capitally done and could not be mistaken.

The disk of the Daisy-flower is worked in a mass of those little knots that sit closely together, the secret of whose making is known to every good needlewoman. They are a capital direct imitation of the group of anthers in the centre of a flower.

The glory of the picture, and what was evidently the delight of the worker, is the Love-in-a-Mist, which stands above the others in the middle top of the picture. The tender blue of the flower, shading to white, the sharply-jagged edges of the petals, the green upstanding forms in the centre, and, above all, the fennel-like divisions of the involucre and the leaves, all lend themselves to satisfactory portrayal with the needle; while the prominent position given to this charming midsummer flower shows how the worker rejoiced in its beauty and took pleasure in painting its form and colour in tender stitchery upon the white silken ground of her picture. The Jasmine flowers, too, are done with evident enjoyment as well as the neat, clear-cut

leaves. The Rose is a Moss-Rose, shown in three stages of bud and half-blown bloom, when this charming Rose is at its best; the mossiness of the calyx being cleverly suggested by short straw-coloured stitches that catch the light upon a ground of dull green. The working material is floss silk, whose silvery, shining surface, dark in some lights, makes a distinct effect of light and shade in the case of the white flowers, even though they are worked upon a ground that is also white.

Sometimes these pictures are of a bunch of flowers without a receptacle, but often there is a basket or vase. In this case there is a basket of very simple form, standing on a darker table worked in the chenilles, which were also much used. They are tiny ropes of silk velvet with an effect of rich short pile, like the old velvets of Genoa.

It is easy to see how the Red Valerian came to be used as a model for needlework. Short stitches and long would easily render the small divisions of the calyx and the long slender spur and single pistil, and a quantity of this, representing the rather crowded flower-head, would have a very good effect on a white or light ground.

The plant itself is a pretty one in any garden. Botanists say that it is not indigenous, but it has taken to the country and acclimatised itself, and now behaves like a native; haunting quarries and railway cuttings in the chalk. It is a capital plant for establishing on or in walls or bold rockwork, as well as in the garden border. It is always thankful for chalk or lime in any form.

Printed in Great Britain
by Amazon

36824469R00119